Economic Instruments for Air Pollution Control

Economy & Environment

VOLUME 9

The titles published in this series are listed at the end of this volume.

Economic Instruments for Air Pollution Control

edited by

Ger Klaassen

International Institute for Applied Systems Analysis,
Laxenburg, Austria

and

Finn R. Førsund

University of Oslo,
Department of Economics,
Oslo, Norway

SPRINGER-SCIENCE+BUSINESS MEDIA, B.V.

INTERNATIONAL INSTITUTE FOR APPLIED SYSTEMS ANALYSIS

LAXENBURG, AUSTRIA

Library of Congress Cataloging-in-Publication Data

Economic instruments for air pollution control / edited by Ger Klaassen and Finn R. Førsund.
p. cm. -- (Economy & environment ; v. 9)
"It consists of a coherent set of papers that were originally presentet at the International Conference on 'Economic Instruments for Air Pollution Control,' held at the International Institute for Applied Systems Analysis (IIASA), Laxenburg, Austria, 18-20 October 1993"--CIP pref.
Includes index.
ISBN 978-0-7923-3151-3 ISBN 978-94-011-1012-9 (eBook)
DOI 10.1007/978-94-011-1012-9
1. Air--Pollution--Economc aspects--Congresses. 2. Air quality management--Cost effectiveness--Congresses. 3. Emission trading--Congresses. 4. Environmental impact charges--Congresses.
I. Klaassen, Ger, 1958- . II. Førsund, Finn R.
III. International Institute for Applied Systems Analysis.
IV. International Conference on "Economic Instruments for Air Pollution Control" (1993 : Laxenburg, Austria) V. Series.
HC79.A4E248 1994
363.73'92--dc20 94-33492
CIP

ISBN 978-0-7923-3151-3

Printed on acid-free paper

The International Institute for Applied Systems Analysis

is an interdisciplinary, nongovernmental research institution founded in 1972 by leading scientific organizations in 12 countries. Situated near Vienna, in the center of Europe, IIASA has been for more than two decades producing valuable scientific research on economic, technological, and environmental issues.

IIASA was one of the first international institutes to systematically study global issues of environment, technology, and development. IIASA's Governing Council states that the Institute's goal is: *to conduct international and interdisciplinary scientific studies to provide timely and relevant information and options, addressing critical issues of global environmental, economic, and social change, for the benefit of the public, the scientific community, and national and international institutions.* Research is organized around three central themes:

- Global Environmental Change;
- Global Economic and Technological Change;
- Systems Methods for the Analysis of Global Issues.

The Institute now has national member organizations in the following countries:

Austria
The Austrian Academy of Sciences

Bulgaria
The National Committee for Applied Systems Analysis and Management

Canada
The Canadian Committee for IIASA

Czech Republic
The Czech Committee for IIASA

Finland
The Finnish Committee for IIASA

Germany
The Association for the Advancement of IIASA

Hungary
The Hungarian Committee for Applied Systems Analysis

Italy
The National Research Council (CNR) and the National Commission for Nuclear and Alternative Energy Sources (ENEA)

Japan
The Japan Committee for IIASA

Kazakhstan
The National Academy of Sciences

Netherlands
The Netherlands Organization for Scientific Research (NWO)

Poland
The Polish Academy of Sciences

Russia
The Russian Academy of Sciences

Slovak Republic
The Center for Strategic Studies

Sweden
The Swedish Council for Planning and Coordination of Research (FRN)

Ukraine
The Ukrainian Academy of Sciences

United States of America
The American Academy of Arts and Sciences

Preface

International negotiations in Europe on a new protocol to control sulfur dioxide emissions, signed in June 1994, have recognized the potential contribution of economic instruments to improve the cost-effectiveness of international agreements. The protocol opens possibilities for countries to jointly implement their emission reduction obligations, although rules and conditions guiding joint implementation have to be elaborated. Such specific rules are necessary since, in contrast to the joint implementation of carbon dioxide emissions (foreseen in the Framework Convention on Climate Change) or chlorofluorocarbon quotas (included in the Montreal Protocol), for emissions of sulfur dioxide the situation is more complex since it is a nonuniformly dispersed pollutant whose environmental impacts depend not only on the total volume emitted but also on the location of the source.

IIASA's Transboundary Air Pollution Project is supporting the negotiations on a new sulfur protocol by means of its integrated assessment model RAINS: the Regional Acidification INformation Simulation model. This model forms an important tool for negotiators for examining cost-effective strategies for reducing acidifying emissions in Europe. In this context, the project also addresses the potential role of economic instruments, among them joint implementation. The basic aim of this part of the research program is to analyze what can be learned from economic theory and methods, as well as from practical experiences, on how economic instruments can be applied, especially for nonuniformly dispersed pollutants, at both a national and international level.

This book is one of the results of this research effort. It consists of a coherent set of papers that were originally presented at the international conference on "Economic Instruments for Air Pollution Control," held at the International Institute for Applied Systems Analysis (IIASA), Laxenburg, Austria, 18–20 October 1993. The papers were selected to provide an overview of results from theory, simulation model studies, as well as practical experiences in the application of economic instruments in both national and international settings. The book should provide interesting reading for economists with an interest in environmental economics, policy makers with a modest background in economics, and, generally, those who are interested in the application of economic instruments.

We appreciate the help of David Pearce (University College, London) in outlining the collection. We are highly indebted to the reviewers who assisted us in reviewing chapters of this book. They needed only little (economic) incentive to come up with helpful comments and suggestions that considerably improved the quality of the book and they therefore deserve special mentioning: Knut Alfsen (Statistics Norway), Scott Barrett (London Business School), Mohammed Belhaj (Göteborg University), Lars Bergman (Stockholm School of Economics), D. Bohi (Resources for the Future), Janusz Cofala (IIASA), Olivier Godard (Centre International de Recherche sur l'Environnement et le Développement), Bente Halvorsen (University of Oslo), Leen Hordijk (Agricultural University Wageningen), Ekko van Ierland (Agricultural University Wageningen), Péter Kaderják (Budapest University of Economics), Snorre Kverndokk (Statistics Norway), Leonid Melnik (Sumy State University), Charles Paulsen (Resources for the Future), David Pearce (University College London), Stef Proost (Catholic University Leuven), Lex de Savornin Lohman (Free University Amsterdam), Thomas Schelling (University of Maryland), Per Schreiner (ECON, Oslo), Karl Steininger (University of Graz), Thomas Sterner (Göteborg University), Tom Tietenberg (Colby College), Harmen Verbruggen (Free University Amsterdam), Hans Vos (DHV Environment and Infrastructure), Doede Wiersma (University of Groningen), Naomi Zeitouni (University of Haifa), and Tomasz Zylicz (Warsaw University).

Furthermore, we gratefully acknowledge the financial support from IIASA not only for basic research but also for organizing the conference and publication of this book. We would also like to thank Margaret Gottsleben for her substantial assistance, Markus Amann and Janusz Cofala for their critical comments and suggestions, Valerie Jones for editing this volume, and IIASA's Publications Department for its support in turning our draft into camera-ready copy. We hope that the book will provide an important contribution to the (joint) implementation of economic instruments and traditional regulations in order to meet environmental goals without squandering scarce natural resources.

Ger Klaassen and Finn R. Førsund
Laxenburg/Oslo
March 1994

Contents

Introduction

Ger Klaassen
Transboundary Air Pollution Project
IIASA, Laxenburg, Austria
Finn R. Førsund
Department of Economics
University of Oslo, Oslo, Norway

Background and Objective

In 1987, the city council of Telluride, Colorado, introduced tradable permits for woodburning. The council required every fireplace or woodburning stove to meet tight emission standards and to have a permit to operate. New stove owners were only allowed to burn wood if they could acquire two permits from existing owners who were willing to sell. As a result, the price per permit rose to over $1,400 and the air pollution problem diminished to the extent that the town now meets the federal air quality standards for particulate matter (Elman *et al.*, 1992). This is just one example of how economic instruments for air pollution control can work in practice.

Economists have a long tradition of advocating the use of economic instruments, such as emission charges or tradable emission permits. Economic theory links the choice of instruments to the fact that pollution is an externality that leads to an inefficient or suboptimal allocation of scarce resources. Consequently, corrective devices are necessary to establish an optimal solution: an outcome that maximizes the benefits or welfare of society. The solution to this problem is straightforward. Polluters need to be confronted with a (Pigovian) tax equal to the marginal external costs or marginal damage of their pollution in order to induce them to take account of the full social costs of their activities (Cropper and Oates, 1992). The classic case for emission charges is that the environmental authority can announce the optimal charge level without knowledge of the source's cost functions, as long as the damage function is linear, and the location of the sources is irrelevant for the degree of damage (Bohm and Russell, 1985). If the sources minimize costs in response to the optimal charge, the charge combines static efficiency and information economy. Instead of setting the Pigovian tax at the optimal level, the environmental agency could also issue permits for emissions, the total number of which equals the optimal emission level, and then allow sources to bid for them. In theory, both quantity and price approaches would yield the same result in terms of efficiency in a setting of perfect information. In spite of the increased willingness to perform contingent valuation studies, however, it is hard to be sanguine about the possibilities to determine optimal emission levels. A more practical approach would be to set environmental quality standards and to achieve these standards at minimum cost. The basic argument in favor of emission charges is that it brings about that allocation of control efforts among polluters that minimizes the costs of pollution control without the costs being known by the authorities. Under the assumption that each source will seek to minimize its costs, it can be proven in a straightforward way that the imposition of a uniform charge per unit of emission will achieve the cost-minimum solution. In the case of imperfect knowledge of pollution control costs, a trial-and-error procedure is needed to meet the environmental goal (Baumol and Oates, 1988). When location starts to matter, this iterative procedure is expected ultimately to meet the goals, but opinions diverge on whether the cost-effective outcome can be attained (Bohm and Russell, 1985; Ermoliev *et al.*, 1993). Similarly, it can be shown that, under a number of restrictive conditions, trading of such emission permits can attain the cost minimum. These conditions are that sources minimize their pollution control costs (i.e., the sum of expenditures on abatement measures and of the net revenues from trading permits), that the permit market is competitive, and that transaction costs are negligible (Montgomery, 1972). This result is obtained irrespective of whether we deal with tradable emission permits or tradable deposition permits.

Evidence from empirical simulation models tends to support the cost-effectiveness claims of economic theory. In an overview of 17 studies, Tietenberg (1985) reports that a shift from regulations (including uniform percentage reductions) to a tradable permit system could cut pollution control costs by 10%–90%. In half of the studies examined the costs savings are more than 40 percent. Welsch (1989), in a study on national uses of tradable sulfur emission permits for power plants in 13 countries in Western Europe, concludes that, depending on the percentage reduction desired, cost savings of 1%–85% are possible.

Practical experiences with the use of economic instruments seem to bear out the Chinese proverb that forecasting is difficult especially if one wants to forecast the future, rather than to confirm the validity of economic theory and simulation models. With a few exceptions (such as lead trading in the USA), the cost savings of emission trading realized in practice have been lower than expected (Hahn and Hester, 1989). The product and emission charges that are applied in practice usually have no or an unknown incentive impact and predominantly serve revenue-raising purposes (Opschoor and Vos, 1989). This is not to say that economic textbooks and model predictions are wrong; the problem is more that economic instruments are rarely applied in the ways envisaged by the models. Economic instruments are usually crafted on top of existing regulatory frameworks instead of being applied in the pure textbook way. High transaction costs, uncertain property rights, as well as regulatory constraints, imply that in reality emissions trading has not achieved the cost savings that are possible under any linear programming optimization model. Charges face similar problems. The additional cost burden of taxes may be opposed by polluters. Environmentalists and regulatory authorities might be hesitant to give too much leeway to polluters (Opschoor and Vos, 1989), or may express concerns about the associated (administrative) costs and enforcement problems involved with the implementation of radical economic incentives. Consequently, charges are hardly ever high enough to induce changes in behavior. Recent developments, however, suggest that we might be experiencing more radical shifts toward the implementation of economic instruments. The passage of the 1990 Clean Air Act amendments in the USA gave way to the trading of sulfur allowances between utilities all over the country. On the other side of the Atlantic, Sweden has implemented high charges on its sulfur dioxide and nitrogen oxides emissions, at levels that are explicitly designed to have an incentive impact. The optimal mix of instruments might be shifting toward the use of economic instruments in a purer form. It is clear, however, that in examining the potential use of economic incentives, it makes sense to combine the results of theoretical and simulation models with an assessment of what happens in practice.

In principle, the analysis of the performance of economic instruments in a national context is fully transferable to the regulation of transboundary pollution in an international context. In contrast to the national case, however, there is no international government that can impose emission charges or establish tradable emission rights. This will require the cooperation of sovereign nations that now can only be expected to cooperate if it will make them better off. The presence of transboundary externalities does imply that gains can be made from cooperative agreements, although there are also incentives for countries not to cooperate (Barrett, 1990; Baumol and Oates, 1988). One fairly general result from economic theory is that, in order to realize the potential welfare gains from the cooperative approach, side payments have to be made to those countries that lose from the full cooperative approach relative to the noncooperative approach (Folmer and Musu, 1992). A second general conclusion is that uniform percentage emission reductions are not only inefficient and too expensive, but in an international context they also create incentives for countries not to participate in such agreements (Hoel, 1992). This is the point where economic instruments can play an important role. As in the national context, emission taxes and emission trading have the advantage that they cut the costs of achieving environmental objectives. Exactly because of this feature, they also increase the chance that more countries will be better off. Although both the tax and the tradable permit systems do possess the desirable property that they can meet a given environmental goal at minimum cost, they might however raise the compliance costs for some countries (Barrett, 1992). In this case, the tax revenues, the initial distribution of permits, or the revenues from selling them, inevitably have to be used as additional instruments in compensating the potential losers in an international agreement.

Again, empirical simulation models lend support to the maintained advantages of economic instruments in an international context. Rose and Stevens (1993), for example, find that the potential gains from worldwide carbon dioxide permit trading are substantial, and that abatement cost savings are on the order of 25%–35%. Moreover, none of the countries/regions they examined appeared to be a net loser. Mäler (1990) analyzes the net benefits from imposing a uniform tax of DM4,000 per ton of sulfur dioxide exported by each country in Europe. He discovers that the aggregate benefits from this export tax are only 1% lower than the total benefits from the full cooperative solution, but stresses that tax revenues need to be used to compensate the losing countries. Klaassen *et al.* (1994) explore the implications of implementing sulfur emissions trading in a European context. They conclude that emissions trading, with a special exchange rate regime, might cut pollution control costs compared to uniform cutback, without, at least in this specific example, violating a given set of deposition goals.

Empirical evidence to support theoretical and modeling results is scarce. This should come as no surprise since the international application of economic instruments has so far been limited. The Montreal Protocol, signed in 1987 to protect the ozone layer, however, is an exception. This Protocol contains provisions that allow countries to transfer their production quotas for chlorofluorocarbons and halons to other parties and enable EC member countries to jointly implement their allocated consumption quotas. Although Bohm (1990) heavily criticizes the Montreal Protocol for its inefficiency, transfers of production quotas actually took place and the relatively small transaction costs suggest a relatively efficient outcome. Furthermore, joint implementation or tradability of agreed emission ceilings is a subject of intense debate not only in the context of the Framework Convention on Climate Change but also in the negotiations on a second sulfur protocol in Europe. In summary, in an international context economic instruments are expected not only to promote efficiency and cost-effectiveness, but also to have attractive distributional consequences. Experience will have to show to what extent these features hold in international practice.

Against this mixed bag of results from theory and simulation models on the one hand, and empirical evidence on the other, the purpose of this book is to give an overview of what can be learned from both economic theory and simulation models, as well practical experience, on the use of economic instruments, in both national and international contexts. The emphasis of the book is on local air pollutants, such as sulfur dioxide, rather than global pollutants such as carbon dioxide or chlorofluorocarbons. Local pollutants are pollutants for which the environmental impacts depend not only on the total volume emitted, but also on the location of the source. The book mainly deals with the use of economic instruments for air pollution control.

Structure of the Book

The book consists of three parts. Part I examines theoretical aspects and the results of simulation modeling of economic instruments in national contexts. Part II focuses on practical, national experiences with economic instruments, and Part III features international applications of economic incentives, particularly emissions trading and joint implementation, and explores theoretical issues, modeling results, and practical design issues.

Part I starts with a review by Atkinson of the theoretical properties as well as the results of empirical simulation models on tradable discharge permits (Chapter 1). Atkinson examines the impact of both constraints on local air quality degradation and long-range transport of acidifying pollutants, as well as the effects of sequential, bilateral trading on the potential cost savings of tradable discharge permit systems. He also discusses the equity features of tradable discharge permit systems. After elucidating the basic theoretical features of both ambient and emission permit systems, a comparison is made with tax strategies. Then the impact of local air quality concerns, long-range externalities and the bilateral, sequential nature of trades on the cost-effectiveness of ambient permit systems in a number of simulation models is analyzed. Finally, the distributional impacts of auctioning and grandfathering tradable permits are examined. Atkinson concludes that constraints on local air quality degradation, as well as on long-range transport can considerably reduce the potential cost savings of both ambient and emission permit systems. Similarly, sequential trading, even with full information, can erode the cost savings of tradable discharge permits. One way out might be the use of ambient permits based on only the most polluted receptor. Auctioning of tradable discharge licenses is expected to reduce their political acceptability since compliance costs increase. Grandfathering appears to be called for to circumvent equity problems.

In Chapter 2, Pototschnig presents a partial equilibrium approach in modeling the use of economic incentives for controlling acidifying sulfur emissions resulting form large combustion plants in the United Kingdom. After sketching the relevant policy background and the arguments for introducing economic incentives, the author goes on to describe the simulation model used for the different sectors: a model simulating the dispatch merit order for the electricity supply industry and a cost minimization approach for refineries and other industries. The results of the simulation exercise are coupled with a description of practical aspects surrounding the use of tradable emission permits. The author concludes that the use of tradable emission permits would be more appropriate than charges. The potential cost savings of introducing tradability of existing emission quotas for only the electricity supply sector would be limited, due to the similarities in the power station mix among the electricity suppliers. Extending the market to refineries and other industries would considerably increase the savings in resource costs.

In Chapter 3, Robinson, Geoghegan and Subramanian leave the narrow realms of a partial equilibrium approach and apply a general equilibrium model to the air quality problem in the Los Angeles basin in order to simulate the direct and indirect economic impacts of emission taxes on air pollution. They start with a discussion of existing regulatory instruments and the new market-based program to control air pollution in the Los Angeles region, and then summarize existing general equilibrium models, specific features of their regional model, and the way in which pollution abatement is treated. The results of imposing emission charges on sulfur dioxide, nitrogen oxides and reactive organic gases on levels of pollution, control costs, and regional value added are described for two cases: with and without technical pollution control measures. The authors find that for levels of emission reductions up to around 30%, total pollution control costs as well as the negative impacts on regional value added, are relatively small. This is so because these levels can be attained with technical abatement measures only. More significant reductions would require much higher emission charges to induce the necessary structural changes in the regional economy.

Part II starts with an overview by Opschoor of the application of economic instruments in OECD countries, especially for air pollution control followed by four chapters containing country-specific case studies.

In Chapter 4, Opschoor describes the current application of economic instruments in OECD countries, and discusses the developments in their use since the publication of the first OECD survey in 1987. After describing the various instruments, Opschoor provides a comprehensive overview of the use of emission and product charges, deposit–refund systems, tradable permits, and enforcement incentives, especially for air pollution. Subsequently, the rationales for using economic and financial instruments are explored, and finally, the observed changes in instrument use are related to changes in policy context. Opschoor concludes that the use of economic instruments, particularly of product charges and deposit–refund systems, has increased since 1987. Half of the product and emission charges employed today are intended to change behavior, although in most cases the data do not allow straightforward conclusions to be drawn on the actual incentive impact. Opschoor concludes that new elements have emerged in discussions on economic instruments, including explicit reference to institutional failure, the relevance of the policy context for instrument choice, awareness of uncertainty, interest in harmonization, the need for pragmatism, and the growing demand for financing environmental measures. The relative role for economic instruments and direct regulation might be changing in favor of the former. It is also clear, however, that economic instruments tend to remain complements to rather than substitutes for traditional regulatory approaches.

In Chapter 5, Lövgren describes and analyzes the rather unique experience in Sweden, with its relatively high emission charges for sulfur and nitrogen oxides and the differentiation of the tax on diesel fuels. First, Lövgren provides an overview of the economic incentives for air pollution currently being used in Sweden against the background of the Swedish energy demand structure and environmental problems. Then, she describes the design of the sulfur tax, its impact on emissions, and its administrative practicability. Subsequently, the nitrogen oxides charge on combustion plant emissions is treated, and the monitoring requirements, administration costs, and the environmental impacts are assessed. The differentiation of the tax on diesel fuels constitutes the final part of her contribution. Lövgren is of the opinion that, although they were combined with traditional forms of regulation, the sulfur tax, the nitrogen oxides charge, and the differentiation of the tax on diesel fuels were the major forces behind the emission reductions observed.

The development of emissions trading for air pollution control in the USA, as an additional asset in the policy makers' ongoing search for an optimal portfolio of instruments, is examined by Kete (Chapter 6). After describing the Clean Air Act, she turns to a discussion of the US experience with economic instruments, especially the emissions trading and marketable permit programs. In doing so she sketches the evolution of the US Environmental Protection Agency's (EPA) emissions trading policy of the 1970s up to the Clean Air Act amendments of 1990. The latter, with its endorsement of sulfur emissions trading between utilities all over the USA, has been one of the most rigorous attempts to apply a tradable permit system. She concludes that the US acid rain sulfur trading program differs fundamentally from the EPA's emissions trading program. Because of its carefully designed institutional structure the acid rain program is expected to lead to a smoothly functioning market in emission permits, while ensuring that the programs goals are met and that both local and regional environmental protection are not affected. This is achieved by combining existing air quality provisions and technology-based (emission) standards with rigorous control mechanisms such as permanent emission caps, explicit emission baselines, continuous emission monitoring, emission tracking, and enforcement penalties well above the marginal control costs.

Chapters 7 and 8 deal with the application of economic instruments in countries in transition to a market economy. The implementation and effectiveness of damage-based air pollution taxes and environmental investment funds in Estonia is the subject of the contribution by Kallaste (Chapter 7). First, he describes the use of economic instruments in environmental policy in Estonia: emission charges, administrative charges, tax differentiation, subsidies, and noncompliance fees. Then he explains how, interestingly, the emission charge for air pollution is based on a weighted average of the volume of the emissions, the size of the polluted area, land use in the polluted area, and the risk to human health of the pollutants emitted. Subsequently, the impacts of these charges on actual firm behavior are examined. Finally, the operation of the Estonian Environmental Fund as the institution that collects and redistributes the revenues from economic instruments is addressed. Kallaste's major conclusion is that the air pollution charges introduced in Estonia were set too low to effectively lead to a reduction in pollution. This low charge level was a political decision to avoid closure of industries. The effectiveness of the system has been further eroded by extremely high levels of inflation and monetary reforms, and is still being undermined by Estonia's slow development to a market economy, implying that elements of the former centrally planned economy, such as soft budget constraints, centrally fixed prices and investments, still exist in many sectors of the economy.

The difficulties associated with the application of economic instruments, especially emission charges for air pollution under conditions of economic reform, are also the subject of the contribution on Russia by Golub and Strukova (Chapter 8). The main question they address is the extent to which the introduction of economic instruments has created incentives for polluters to alter their behavior. They start by describing the command-and-control system in place before the introduction of economic incentives. Subsequently, they describe the introduction of an experiment with emission fees based on damage estimates or on the level of expenditures necessary for pollution control, and the adoption and recent modification of these fees for the whole of Russia. Finally, they estimate the impacts of the fee system on expenditures for environmental protection and on emissions. They conclude that although the command-and-control system was unable to solve many environmental problems in the former Soviet Union, to some degree it did prevent further deterioration. The introduction of fees did lead to an increase in the number of enterprises with agreed emission levels and standards, but in spite of economic decline, the pollution intensity per unit of GNP has increased and the share of investments in pollution control abatement in GNP has dropped. The environmental management system and the fees have not been very effective in protecting the environment. The fees are not high enough to provide the revenues necessary for environmental protection because in many cases the fees could not be collected due to the insolvency of many enterprises. Furthermore, the fees do not appear to have created incentives for firms to invest in pollution control equipment, and in most cases have led to an increase in output prices rather than increased environmental investments.

Part III contains four studies dealing with international aspects of the implementation of economic instruments. The efficiency gains and distributional implications of joint implementation and tradable emission quotas for controlling greenhouse gases are examined by Bohm in Chapter 9. After a comparison of the pros and cons of different carbon taxes compared to quantity controls, Bohm asks whether joint implementation can be efficient, monitored reliably, and can evolve into a system of tradable emission quotas. Subsequently, he discusses the impact of abatement costs and increased fossil fuel use on the distributional implications of quota tradability for both developing and industrialized countries. In conclusion, Bohm maintains that a serious drawback of joint implementation is the difficulty of estimating the net emission reduction from individual projects. Joint implementation does seem to be politically acceptable and appears to be an easy way to start, although it is doubtful whether this will lead to a significant increase in efficiency. In contrast to joint implementation, Bohm expects quota tradability to improve efficiency, especially if developing countries could be encouraged to participate.

In a theoretical contribution, Nentjes analyzes the extent to which various motives influence the effects of financial transfers on the welfare gains of parties to international agreements (Chapter 10). Nentjes starts with a survey of the essential concepts of a cooperative and noncooperative Nash–Cournot equilibrium and market equilibrium in the context of a two-country model of reciprocal transboundary pollution. Then he replaces the standard assumption that countries are only interested in their own costs and level of pollution, with the alternative assumptions that some countries are also concerned about the environmental quality in other countries. The question now is to what extent this alternative assumption leads to a modification of the standard results. Nentjes proceeds to show that international care can lead to a change in the rate of exchange of reciprocal emission reduction commitments in favor of the other country. Alternatively, as a joint implementation project, the country that cares can pay for taking additional pollution control abroad, while simultaneously increasing its own emissions.

Practical aspects and simulation models of controlling sulfur emissions in Europe are dealt with in Chapters 11 and 12. Førsund and Naevdal model sulfur emissions trading in Europe and discuss possible administrative frameworks within which this could take place (Chapter 11). Against the background of the negotiations on a new sulfur protocol in Europe, the authors first review the theory of emissions trading for regional pollutants, such as sulfur, where the location of the source is relevant. They then discuss the conditions necessary for making emissions trading between parties to the protocol worthwhile, and propose the use of exchange rates to guide bilateral trades. Subsequently, centralized and decentralized administrative frameworks for dealing with trades are described. Finally, they describe the results of a model that simulates emissions trading as a bilateral, sequential process under various assumptions on the initial distribution of emission reduction commitments. In their model, trading takes place according to exchange rates to account for the fact that sulfur emissions disperse nonuniformly. They conclude that agreement on environmental objectives and the minimum-cost solution is needed for the implementation of their system of emissions trading with fixed exchange rates. Their model results suggest that the exchange rate emissions trading they propose would be especially attractive from a cost-effectiveness point of view if the initial distribution of emissions does not deviate too much from the optimum solution.

In Chapter 12, Burtraw seeks to analyze whether countries or individual companies should be trading sulfur emission permits in Europe and what the impact would be on allocative efficiency and cost-effectiveness (productive efficiency). First, Burtraw compares the productive efficiency of allowing countries or individual firms to trade. After a short discourse on the role of spatial considerations, the author reports on the simulation of sulfur emissions trading in Europe exploring the performance of bilateral, sequential trading by both individual enterprises and national governments. Last, but not least, the impact of the choice of trading agents on allocative efficiency is examined. Burtraw concludes that it is not obvious whether cost-effectiveness would be achieved if individual firms rather than countries were to be the trading agents, although placing permits in the hands of individual enterprises would ensure that incentives to cut costs would remain in place. His simulation model results confirm that trading between enterprises outperforms trading between national governments. He also finds it to be unlikely to be rational for individual countries to adopt an incentive-based approach that promotes allocative efficiency, since this might result in disadvantages in international trade.

Concluding Remarks

The chapters collected in this book suggest the following conclusions. Constraints on local air quality and long-range transport of pollutants, the sequential character of trades, and the restriction of trade to sectors with small differences in marginal costs, might considerably reduce the expected cost savings from tradable discharge permits. To satisfy equity concerns, grandfathering of tradable permits seems warranted. As long as environmental objectives can be attained with technical means at relatively low cost, general equilibrium or macroeconomic impacts also tend to be small, since structural changes hardly occur.

The practical application of economic instruments in OECD countries, especially the use of product and emission charges, has increased. Moreover, more than in the past, these charges are explicitly designed to have an incentive impact although the actual incentive impacts are still largely uncertain. In applying economic instruments, there is a need for pragmatism and greater attention to the policy context in which they are applied. It is increasingly being recognized that economic incentives are complementary to rather than substitutes for regulation. The Swedish experience with sulfur and nitrogen oxides charges shows that charges, combined with regulation, can be set so as to achieve the desired environmental impacts. The experiences in Russia and Estonia, however, suggest that pollution taxes may fail to affect behavior, especially in countries in transition due to the lack of political will to set charges at levels that may lead to the closure of firms, hyperinflation and monetary reforms, as well as the lack of competition and the continued prevalence of centralized planning in major industrial sectors. Even in the US market economy, radical reforms in the application of emissions trading under the 1990 Clean Air Act, in order to ensure fluid permit markets, are combined with command-and-control elements such as air quality and emission standards and rigorous monitoring and enforcement.

In the international context, trading of carbon quotas holds the promise of being more efficient than joint implementation, especially if developing countries can be encouraged to participate. Concern for the environment in other countries forms an additional incentive for joint implementation of emission reductions but care for other countries' pollution control costs might have the opposite impact. Sulfur emissions trading in Europe can be used to promote the cost-effectiveness of a new sulfur protocol but requires agreement on the objectives and the use of exchange rates reflecting the atmospheric dispersion of sulfur. Trades can proceed decentralized or centralized, but trading between individual firms is expected to outperform trading between national governments in terms of cost-effectiveness. As with the national application of economic instruments, learning-by-doing seems to be required at the international level, to find both the optimal mix of instruments and the most appropriate institutional design.

References

Barret, S., 1990, The Problem of Global Environmental Protection, *Oxford Review of Economic Policy* 6(1):68–79.

Barret, S., 1992, *Convention on Climate Change: Economic Aspects of Negotiations*, OECD, Paris, France.

Baumol, W., and W.E. Oates, 1988, *The Theory of Environmental Policy*, Cambridge University Press, Cambridge, UK.

Bohm, P., 1990, *Efficiency Issues and the Montreal Protocol on CFCs*, Environment Working Paper No. 40, World Bank, Washington, DC, USA.

Bohm, P., and Russel, C.S., 1985, Comparative Analysis of Alternative Policy Instruments, in A.V. Kneese and J.L. Sweeney, eds., *Handbook of Natural Resource and Energy Economics*, Vol. 1, Elsevier Science Publishers, Amsterdam, The Netherlands.

Cropper, M.L., and W.E. Oates, 1992, Environmental Economics: A Survey, *Journal of Economic Literature* **30**:675–740.

Elman, B., Tyler, T., and Doonan, M., 1992, *Economic Incentives Under the New Clean Air Act*, Paper presented at the 85th Annual Meeting of the Air and Waste Management Association, Kansas City, Missouri, 21–26 June 1992, EPA, Washington, DC, USA.

Ermoliev, Y., Klaassen, G., and Nentjes, A., 1993, *Incomplete Information and the Cost-efficiency of Ambient Charges*, WP-93-72, International Institute for Applied Systems Analysis, Laxenburg, Austria.

Folmer, H., and Musu, I., 1992, Transboundary Pollution Problems, Environmental Policy and International Cooperation: An Introduction, *Environmental and Resource Economics* **2**(2):107–116.

Hahn, R.W., and Hester, G.L., 1989, Marketable Permits: Lessons for Theory and Practice, *Ecology Law Quarterly* **16**:361–406.

Hoel, M., 1992, International Environment Conventions: The Case of Uniform Reductions of Emissions, *Environmental and Resource Economics* **2**(2):141–159.

Klaassen, G., Førsund, F., and Amann, M., 1994, Emission Trading in Europe with an Exchange Rate, *Environmental and Resource Economics* **3**:1–26.

Mäler, K.G., 1990, International Environmental Problems, *Oxford Review of Economic Policy* **6**(1):80–108.

Montgomery, W.D., 1972, Markets in Licenses and Efficient Pollution Control Programs, *Journal of Economic Theory* **5**:395–418.

Opschoor, J.B, and Vos, H., 1989, *Economic Instruments for Environmental Protection,* OECD, Paris, France.

Rose, A., and Stevens, B., 1993, The Efficiency and Equity of Marketable Permits for CO_2 Emissions, *Resource and Energy Economics* **15**:117–146.

Tietenberg, T.H., 1985, *Emissions Trading: An Exercise in Reforming Pollution Policy*, Resources for the Future, Washington, DC, USA.

Welsch, H., 1989, Kosten der SO_2-Minderung unter alternativen umweltpolitischen Strategien: Ergebnisse eines Simulationsmodells des westeuropäischen Kraftwerksektors, *Zeitschrift für Energie und Wirtschaft*, 1/89, pp. 51–59.

Part I

National Model Studies

Chapter 1

Tradable Discharge Permits: Restrictions on Least-Cost Solutions

Scott E. Atkinson
Department of Economics
University of Georgia, Athens, GA, USA

Abstract

Conceptually, tradable discharge permit systems have the potential to yield the least-cost solution to achieve ambient standards. Empirical evidence indicates that command-and-control solutions are substantially less cost-effective. Despite this, few emissions trades have been made in the USA or Europe. This paper attempts to explain this phenomenon by measuring the substantial reductions in potential cost savings due to restrictions on local environmental degradation, the generation of externalities, and the nature of real-world trading of permits. We model the effects of zonal trading, emissions trading rules, and limits to acid deposition, as well as the bilateral and sequential nature of actual emissions trades. We also examine compromise systems that recoup some of these losses. Schemes to optimally sequence trades capture only some of the potential cost savings of an ambient permit system. However, an ambient permit system based on the single worst receptor preserves most of the potential cost savings of a full ambient permit system. For any of these systems, permits will most likely need to be grandfathered to satisfy equity concerns.

Key words: tradable discharge permits, least-cost solutions, ambient permits, emission permits, externalities, long-range acid deposition, emissions trading rules, zonal trading, sequential and bilateral trading, equity.

1.1 Introduction

Since the early 1970s, economists have modeled the costs of tradable discharge permit (TDP) systems relative to those of the command-and-control system for achieving air quality standards. Under these systems, the pollution control authority is required to know the optimal number of permits to issue, and how individual source emissions map into ambient air quality, but does not need to know individual source marginal control costs. Sources could be required to purchase either of two types of permits for each pollutant that it discharges. Emission permits (EPs) could be required for each unit of uncontrolled emissions, or ambient permits (APs) could be required for each unit of ambient degradation at one or more receptors. High marginal cost of control sources purchase permits from low marginal cost of control sources, which actually control emissions.

Proofs of the existence of a competitive least-cost equilibrium for an ambient permit system have been provided by Baumol and Oates (1971) and Montgomery (1972). This system yields least-cost regional emission control for achieving a given level of air quality measured at regional receptors. Each source must purchase APs from each receptor that the source affects, if the receptor registers ambient readings equal to the standard. Considerable empirical evidence from mathematical programming simulations, much of which is summarized in Tietenberg (1985), indicates that savings from adopting an AP system is substantial relative to the command-and-control solution.

In 1975 the US Environmental Protection Agency (EPA) initiated the Emissions Trading Program, which was designed to allow the maximum cost-reducing trading activity while maintaining local air quality standards and limiting long-range pollutant deposition. The control authority can allow trading of EPs but must ensure "nondegradation", i.e., that national ambient standards are satisfied and that aggregate emissions do not increase. To enforce national ambient standards, some states require that emissions trades result in a net reduction in emissions; some employ trading within small zones to minimize "hot spots" (which are a violation of the ambient standard in a region that was formerly in compliance); and others require dispersion modeling. For nonuniformly mixed pollutants such as sulfur dioxide (SO_2), nitrogen oxides, and particulates, air quality is a function of the level of emissions and their location. Therefore, whenever trades involve discharge points with different stack heights or locations, increased degradation of air quality is possible. Emission reduction credits are issued to sources which overcontrol emissions. These credits either can be used by a firm to satisfy the standard at another discharge point or can be sold.

While cost savings have been smaller and trades fewer than expected under the Emissions Trading Program, trading activity may increase due to recent legislation. The Wisconsin Fox River TDP program, as described in Hahn (1989), produced only one trade and minimal cost savings. Only about half a dozen firms negotiated in each of two trading clusters, so that markets were thin. Further, only pulp and paper mills were involved so that marginal costs may have been too similar and pressures not to participate may have been substantial. In nonattainment areas (where ambient air quality standards are exceeded), while more trades have occurred between different sources within a given firm, fewer inter-firm trades have been observed. Some inter-firm trading has been observed in the tetraethyl lead trading program of 1982–1987 (Hahn, 1989). While refineries traded and banked lead rights to achieve estimated cost savings of $228 million, participation was limited in terms of geography and type of firm.

Recent acid rain legislation may stimulate more emission trades. To reduce acid rain, the Clean Air Act (CAA) Amendments of 1990 (Public Law 101-549) required a reduction in SO_2 emissions by 10 million tons (US Congress, 1990). This is to be implemented by giving SO_2 EPs to electric utilities, which can be bought and sold nationwide or banked for future use, thereby extending earlier efforts by the EPA to introduce the "bubble" and "offset" policies. The former allows an existing emitter to relax the control on one or more sources of a particular pollutant, provided that it secures an equivalent reduction of the same pollutant from some other nearby source. The latter allows new sources of pollution to enter a region where ambient standards are already exceeded, provided that the new source procures sufficient reductions in emissions from existing sources (above their previously mandated levels) to guarantee that the air quality will be improved as a result of the transaction.

Many characteristics of TDPs may have prevented their potential cost savings from being realized, as discussed in Howe (1993). He compares TDP systems versus taxes in terms of geographic applicability, predictability of outcome, minimum cost achievement of standards, revenue generation, dynamic adaptability to changing pollution demands, incentives to adopt innovative pollution technology, equity, bias against new entry, and possible causes of inefficiency.

In this paper, while I also discuss the equity features of TDPs, I differ from Howe by concentrating on the reduction in potential cost savings of TDP systems due to: (1) constraints on local or external environmental degradation, and (2) the bilateral and sequential nature of real-world trading of TDPs. In the process of minimizing control costs, the AP system increases emissions relative to the EP and, typically, the command-and-control systems. Further, the AP system creates strong incentives to build taller stacks for the release of pollutants, which results in externalities in the form of inter-country transport of pollutants. However, the Emissions Trading Program disallows trades that increase emissions, and seeks to reduce long-range acid deposition. Restrictions on the AP and EP systems can involve the use of trading zones within local airsheds and countries, emission trading rules which employ constraints on total emissions and local ambient degradation, and EPs for long-range deposition. We estimate the effects of these restrictions on the potential cost savings of TDPs. An ambient permit system based on only the most degraded air quality receptor appears to hold promise as an administratively simple system that captures most of the cost savings of the AP system. However, it shares the other fundamental characteristics of the AP system. Pollution permits are typically traded in a sequential and bilateral manner. Recent research indicates that the gains from such trade will fall considerably short of the savings that could be obtained from instantaneous trading. A number of full and limited-information trading systems are considered. Finally, since substantial resistance to TDPs has also been raised on equity grounds, the financial burden of auctioning and "grandfathering" of TDPs is also considered.

The remainder of this paper is organized as follows. In Section 1.2 we first review the formulation of the AP and EP systems and summarize their basic theoretical and empirical properties. Section 1.3 adds an additional perspective on TDPs by drawing a comparison with tax strategies. Section 1.4 focuses on the role of restrictions on local degradation, externalities, and bilateral and sequential trading in reducing the least-cost potential of the AP system. The equity of alternative systems for distributing TDPs is addressed in Section 1.5. A summary and suggested modifications of TDPs follows in Section 1.6.

1.2 Pure Ambient and Emission-Based TDPs

Considerable linear programming effort has been devoted to modeling AP and EP systems. Most of this effort has focused on two alternative approaches, differing in the structure of the constraint relationships. The AP model employs individual-source marginal control costs and individual emission dispersion characteristics to compute the allowable source emissions which will achieve ambient air quality standards at least cost. The EP model also uses individual-source marginal control costs, but assumes that a unit of emissions will have the same impact on ambient air quality regardless of the source.

Although any nonlinear convex control cost function can be assumed for the firm, we employ a quadratic function for simplicity. Then the ambient least-cost (ALC) problem is the solution to the following quadratic programming problem:

$$\min \quad z = \sum_j c_j x_j + \sum d_j x_j^2, \tag{1.1}$$

$$\text{subject to} \qquad \sum_j a_{ij} x_j \geq b_i, \qquad i = 1, \ldots, m, \tag{1.2}$$

$$x_j \geq 0, \quad j = 1, \ldots, n, \tag{1.3}$$

where b_i is the reduction in particulate concentration required to achieve the standard at the ith receptor ($i = 1, \ldots, m$); c_j, d_j are coefficients representing the cost of control per day for the jth source ($j = 1, \ldots, n$); x_j is the number of tons to be removed per day by the jth source; and a_{ij} is the transfer coefficient which relates emissions from the jth source to air quality at the ith receptor.

Equation (1.2) guarantees that reductions in ambient concentrations will be at least sufficient to meet the standard at receptor i. The shadow prices which solve the dual to the ALC problem become permit prices in the AP system. Firms must purchase a permit for each receptor where they produce ambient degradation, as long as that receptor is degraded to the level of the ambient standard, i.e., the corresponding constraint is "binding".

Ambient air quality standard q implies a unique b_i^q ($q = 1, \ldots, t$) at each receptor. The volume of ambient permits to be issued at receptor i for standard q is

$$p_i^q = u_i - b_i^q, \qquad i = 1, \ldots, m, \tag{1.4}$$

where u_i is uncontrolled air quality at receptor i; and b_i^q is required improvement in air quality at receptor i for standard q.

Assuming that costs are a quadratic function of emission control, we can write the emissions least-cost (ELC) problem as the solution to a quadratic programming problem:

$$\min \ z \tag{1.5}$$

$$\text{subject to} \quad \sum_j x_j \geq e, \tag{1.6}$$

and (1.3), where e is the aggregate amount of regional emissions that must be removed to meet ambient standards at the region's worst air quality receptor.

For the ELC solution to be comparable to the ALC solution, the x_j from the ELC solution must satisfy (1.2) with at least one equality. Unless pollutants are uniformly mixed, it will not be known *a priori* whether a given e will be sufficient to accomplish this, since (1.2) is not part of the ELC solution. If no constraint holds with equality, then e must be scaled down. If air quality standards are violated, e must be scaled up.

There are many similarities between the AP and EP formulations and their implicit assumptions. The objective function to be minimized in (1.1) and the constraints in (1.3) are the same for both permit strategies. All sources have a dynamic incentive to lower their marginal costs of control through R&D. Both formulations assume competitive markets, the independence of individual source control functions, rule out synergistic effects among pollutants or individual source emissions, and assume linearities in the transfer coefficients.

Important differences exist between the two systems as well. Under the AP system, the control authority only has to decide on how many permits to issue, as defined by (1.4). It is not required to know individual source marginal control costs or transfer coefficients. The competitive market produces the primal solution by generating permit prices as defined in the dual solution. In the EP system, since the constraint inequality in (1.2) is replaced by (1.5), a uniform market price obtains for all EPs. To satisfy (1.2) with at least one equality, the control authority will most likely have to adjust the volume of permits over time as they are traded among sources.

Atkinson and Lewis (1974, 1976) examined the costs of control of the 27 largest point source emitters of particulate matter in the St. Louis region for the EP and AP systems. Costs under the AP strategy are substantially less than those under the EP strategy, because of its simplifying assumption that emissions from all sources have the same impact on air quality. Therefore, the EP model describes the least costly control strategy to achieve a required reduction in regional mass emissions, but not necessarily a required level of ambient air quality, which is achieved by the AP system. The EP system results in substantial overcontrol of more distant sources or ones with taller stacks, thus increasing the cost to achieve specified air quality levels. Individual states in the USA are required to prepare so-called "command-and-control" air quality strategies, which are State Implementation Plans (SIPs) that provide detailed control requirements to achieve federal ambient standards. The SIP strategy which we modeled, representative of state SIP strategies, is far more costly than either of the TDP strategies over a wide range of air quality. This is because both individual marginal control costs and emission dispersion characteristics are ignored in developing the SIP strategy. The SIP strategy is 6–10 times more costly than the AP strategy, and 1.3–6 times more costly than the EP strategy, which enjoys one-half of the cost savings of the AP strategy. These results are consistent with those of McGartland and Oates (1985) and others cited in Tietenberg (1985).

The AP system has the potential to become administratively complex. To ensure the cost-effective allocation of control responsibility among emitters subject to attainment of ambient standards at all receptor locations, the control authority would have to create separate permit markets for each pollutant and receptor at which air quality is degraded to the level of the ambient standard [that is, the constraint is binding in (1.2)]. While few constraints are binding at the primary standard (designed to protect human health), nearly all are binding at the secondary standard (designed to protect human health and welfare). The EP system is simpler, since only one permit market is required for each pollutant.

A second potential problem with the AP system is that it meets ambient air quality standards at each receptor at a lower total cost than the EP system by allowing more emissions and by more uniformly degrading ambient air quality. While total emissions are clearly of secondary importance under the Clean Air Act, increases in emissions through trades have not been allowed by the courts.

A potential dynamic problem with the EP system is that it may fail to meet ambient standards as new sources enter a region. This system guarantees that total emissions will remain unchanged as new sources enter a region, since each new source must purchase EPs from existing sources (offsets). If, however, the permits are purchased from emitters in relatively clean parts of the region and are used in relatively polluted parts of the region, hot spots may result.

1.3 Alternatives: Tax Strategies

With limited information, the AP strategy appears to be superior to tax strategies. Without knowledge of source marginal control costs and transfer coefficients, control authorities would have to impose an *ad hoc* uniform emissions tax. Both an EP system and a uniform emissions tax would have to be revised iteratively to meet air quality standards, whereas to implement the AP system, the control authority only needs to decide on the number of permits to issue. Iterative adjustment of this number and knowledge of individual control costs or transfer coefficients are not required. However, tax and permit systems share the same desirable characteristics that cost-minimizing decisions would be made internally by the firm and plants are given the incentive to adopt the most cost-effective control technology. In addition, the administrative costs of both TDP systems and the tax system should be less than those of the SIP system.

A second tax strategy involves levying taxes, determined from the ALC or ELC dual algorithms, on all source emissions. The information requirements are obviously far greater than with the TDP systems. The set of ALC (ELC) taxes achieves the required ambient air quality (emission reduction) at minimum cost, while taxing in relation to environmental degradation. An ALC-based tax would imply different emission taxes for different sources, since in the ALC dual solution shadow values differ across receptors and transfer coefficients differ across sources. An ELC-based tax would imply a uniform emissions tax for all sources, since a single, worst-air-quality receptor is employed and all source emissions are assumed to be uniformly dispersed. These taxes reproduce the ALC and ELC primal solutions. Sources required to control in the primal solution are taxed at a rate equal to the marginal cost of control in the minimum-cost solution, and those not required to control are taxed at a rate less than marginal cost (to raise tax revenues) or at a zero rate (if no additional revenues are desired). Taxing all emissions provides dynamic incentives for all sources to undertake R&D to lower the marginal costs of control. The tax rate itself will decline for a given air quality improvement if lower marginal cost control devices are adopted.

Private compliance costs comprise control costs and emissions taxes (or permit charges), the latter of which are not social costs. However, large compliance costs may appear inequitable and make the implementation of a strategy infeasible politically. As an example from Atkinson and Lewis (1976), under the ALC-based tax strategy, compliance costs per year at the primary standard are approximately $303,750, comprising $66,332 in control costs and \$237,418 in taxes on uncontrolled emissions, compared with \$66,332 under the AP system. However, this tax strategy is less expensive than the EP system costs of \$306,859, and substantially less than the SIP strategy costs of almost \$3,000,000.

1.4 Modified TDPs

We now examine a number of constraints to the AP and EP solutions which reduce their cost savings. First, we hold emissions and air quality constant by employing trading zones, constraints on long-range deposition, or emissions trading rules. Then, we attempt to model the sequential and bilateral nature of TDP trading.

1.4.1 Zonal TDP systems with emissions constraints

Atkinson and Tietenberg (1982) examined how TDP markets that minimize the costs of achieving air quality standards can be simplified to reduce the administrative burden and still capture most of the cost savings of the AP and EP systems. A highly simplified version of the AP system, the highest ambient permit (HAP) system, defines a single permit market in terms of the most degraded receptor. This system can be written as:

$$\min \quad z \tag{1.7}$$

$$\text{subject to} \qquad \sum_j a_j^* x_j \geq b^* \tag{1.8}$$

and (1.3), where b^* is the greatest required reduction in pollution concentration among the i receptors; and a_j^* is the transfer coefficients relating emissions from source j to ambient concentrations at the receptor with the greatest required reduction in pollution concentration.

Three multiple-zone modified EP systems were also examined which allow trades only within zones. They are characterized by the same programming problem:

$$\min \quad z \tag{1.9}$$

$$\text{subject to} \qquad \sum_j \delta_{jw} x_j \geq e_w, \qquad w = 1, \ldots, r \tag{1.10}$$

and (1.3), where e_w is the aggregate required emissions reduction in zone w; and δ_{jw} equals 1 if source j is in zone w, and 0 otherwise. The value of

e_w is determined as the sum of SIP allowed emissions within each zone or computed using an ambient rollback calculation.

The HAP system achieves the desired air quality goals at a cost only slightly higher than the AP system. While an influx of new sources could cause a "hot spot" problem, the probability that firms will relocate in response to environmental costs is small. Total emissions will increase relative to the SIP strategy.

We also examine systems that divide the region into zones to limit the hot spot problem. The first is a system of uniform zones within which EPs can be traded. However, this uniform zonal discharge permit (UZDP) system increases compliance costs and fails to consistently reduce the possibility of hot spots. We also consider larger zonal discharge permit systems, each of whose zones exceed those under the UZDP system. For most *ex ante* air quality levels, at least one of the larger zone systems has a lower reading at the worst receptor than the UZDP system and yields cost savings between those of the UZDP and HAP systems. However, total control costs are considerably lower than under the SIP system.

All of the permit systems examined except the AP system allow air quality at some receptors to exceed the standard. Hence, all permit systems other than this system will require some safety margin to ensure compliance at all receptors.

1.4.2 TDP Systems with constraints on externalities

The dramatic increase in acid rain deposition in the northeastern USA and large parts of Canada over the past few decades has been well documented. Although most of this increase has been traced to Midwestern sources, it is not primarily due to increased levels of SO_2 emissions, as was once widely thought; these emissions have remained relatively constant. Instead, one major cause appear to be the more complete oxidation of SO_2 into sulfates (SO_4) and hence acid rain, through the greater use of coal to produce electricity during summer months. A second major cause is an approximately threefold increase in the height of emission stacks, which results in a greater percentage of SO_2 emissions traveling longer distances, greater conversion of SO_2 to SO_4 in transit, and therefore greater long-range acid deposition. Under the pre-1990 system of controls based on SIPs, most Midwestern sources were allowed to increase stack heights in lieu of installing SO_2 flue gas desulfurization equipment (popularly known as "scrubbers"). Increased stack heights sufficiently reduced ambient SO_2 concentrations at ground level to satisfy federal ambient standards in a local airshed.

From Atkinson (1983), greater long-range acid deposition is typically created by a local AP system than by a local EP system, [ince total emissions per unit of ground-level concentration of the pollutant is at least as great under the former. Further, more emissions are released at a greater stack height and more emissions are transported from the region under the AP system. If local cost savings under this system are substantial, political pressure for its adoption may be great. However, increased long-range acid deposition may lead to a nonoptimal global solution. Further, when externalities are internalized under a global AP system by limiting the amount of long-range acid deposition, the cost advantage of this system may be reduced sufficiently to make its adoption nonoptimal relative to a global SIP system.

To determine the magnitude of control cost savings as well as increased long-range acid deposition under these local and global systems, Atkinson (1983) solved a nonlinear program using data on SO_2 control costs and transfer coefficients for major point sources in the Cuyahoga County region of the Ohio River Basin. This region is presumed to contribute significantly to acid deposition in Canada and the northeastern USA. Local and global solutions for the SIP strategy and both EP and AP markets were simulated. The local solutions were obtained using the quadratic programs that yielded the previous AP and EP solutions. The global solutions were obtained by adding the following constraint on long-range SO_4 transport, which reduces acid deposition from all sources by a prescribed amount:

$$\sum_v a_v \sum_j x_j^k \geq r, \tag{1.11}$$

where a_v is the long-range transport coefficient mapping SO_2 emissions from Cuyahoga County into $\mu g/m^3$ of SO_4 in the vth neighboring region; r is the required reduction in SO_4 concentrations in all regions due to Cuyahoga County SO_2 emissions; and x_j^k is the number of tons to be removed per day by the jth source under the kth strategy (k = SIP, ELC or ALC).

Considering local strategies, the EP strategy is far more costly and the AP strategy is far less costly than the SIP strategy to achieve local ambient SO_2 standards. However, the local AP system loads the local environment more heavily with SO_2 emissions at tall stack heights, relative to the current SIP system. This substantially increase the long-range acid deposition by the local AP system. When constraints limiting acid deposition are introduced to achieve global solutions, we observe a significant reduction in the cost savings of the AP relative to the EP and SIP strategies. The remaining cost savings may be insufficient to offset the increased transactions costs and administrative costs associated with implementing the global AP strategy.

The magnitude of environmental loading under the local AP system is consistent with results of Atkinson and Lewis (1974) and Seskin *et al.* (1983). The former find for St. Louis that the AP system removes 15% (80%) of the particulate matter controlled by the SIP system and 20% (70%) of that removed by the EP system at the primary (secondary) air quality standard.

1.4.3 Emissions trading rules

Since the CAA has been interpreted to rule out emissions trades that increase total emissions or degrade ambient air quality, Atkinson and Tietenberg (1987) examined emissions trading rules that satisfy this requirement. The *nondegradation offset* trading rule is modeled by adding the constraint

$$\sum_j x_j \geq s^* \tag{1.12}$$

to the problem in (1.1)–(1.3), where s^* is the minimum amount (in tons) by which emissions must be reduced regionally as determined by the required total emissions reduction under the SIP system. This system guarantees that air quality constraints are satisfied and that total regional emissions do not increase.

The *modified pollution offset system* allows trades so long as neither the pre-trade air quality nor the ambient standards are violated. No constraint is placed on total emissions. This system combines (1.1) and (1.3), with

$$\sum_j a_{ij} x_j \geq \max\{b_i, b_i^*\}, \quad i = 1, \ldots, n, \tag{1.13}$$

where b_i^* is the required reduction in pollution concentration at receptor i to achieve the level of air quality under the current regulatory system.

These two trading rules are alternatives to the pollution offset system proposed by Krupnick *et al.* (1983), which requires that trades cannot violate ambient standards. Our analysis suggests that no single "second-best" trading rule dominates in all circumstances. The modified pollution offset can protect the air quality at specific receptors while facilitating trades, but it typically does so by increasing emissions from tall stacks. Although the nondegradation offset system has been shown to result in lower control costs while securing a greater reduction in emissions, this result has also been shown to be quite sensitive to the nature of the SIP system. In particular, if we base the SIP control responsibility for achieving the ambient standards on the proportion of uncontrolled emissions from each source before trade is allowed, then the modified pollution offset system obtains a cost advantage. The use of either trading rule results in substantial cost reductions relative to the SIP strategy.

1.4.4 Sequential trading

While the bubble policy has resulted in significant cost savings, it has consumed only a small fraction of the trades and cost reductions that appear to be possible from mathematical modeling of the AP and EP systems. Atkinson and Tietenberg (1991) attribute this divergence for nonuniformly mixed pollutants to the nature of the actual trading process, which is sequential and bilateral and, hence, differs considerably from the implicitly simultaneous trading process modeled in mathematical programming studies.

To explore this issue we employed an algorithm that mimics the essential elements of the trading process as it actually occurs under the bubble policy. A full-information, sequential-trading scenario is used to model bilateral trading where cost savings are maximized. We first create a matrix of the cost savings from each possible pairwise feasible trade, assuming that only this trade takes place. All trades are assumed to equalize the marginal control costs of the two trading firms.

Assume that the fitted total cost function for the jth source is

$$\widehat{C}_j = \hat{\beta}_j x_j^2, \quad j = 1, \ldots, n, \tag{1.14}$$

where $\hat{\beta}_j$ is a fitted coefficient. Then fitted marginal cost is

$$\widehat{mc}_j = 2\hat{\beta}_j x_j, \quad j = 1, \ldots, n. \tag{1.15}$$

Assume that source 2 has the lower marginal cost, so that

$$\hat{\beta}_1 x_1 > \hat{\beta}_2 x_2. \tag{1.16}$$

By making (1.13) hold as an equality, we determine the number of permits, T, that source 2 would sell to source 1 as

$$\hat{\beta}_1(x_1 - T) = \hat{\beta}_2(x_2 + T), \tag{1.17}$$

and solve for T as

$$T = (\hat{\beta}_1 x_1 - \hat{\beta}_2 x_2)/(\hat{\beta}_1 + \hat{\beta}_2), \tag{1.18}$$

where the numerator is positive, from (1.13).

A feasible trade is one that assures that, pre- and post-trade, the ambient standard is met. The trade producing the largest cost savings is selected and assumed to be consummated. The emission and air quality vectors are updated to reflect the post-trade emission patterns, and the cost savings are recorded. The matrix is then recalculated for the remaining sources (eliminating the row and column corresponding to the two trading partners who have just traded). This recalculation is required since post-trade air qualities have changed due to the rearrangement of emissions from the trade, which changes the set of future feasible trades. This process is iterated until the last feasible trade has been made.

Two limited-information scenarios are also considered. In the first, trades are sequenced by selecting the firm with the lowest marginal cost as the seller of permits in each iteration, the best set of feasible trades is found, and trading is consummated as in the full-information process. In the second limited-information scenario, the seller in each iteration is selected randomly and then trades are determined as with the other scenarios.

Again using St. Louis data, the sequential, full-information system achieves only 88% (50%) of the potential control costs savings of the AP system relative to the SIP system to achieve the primary (secondary) standard. The limited-information systems typically achieve less than one-half of these savings. Thus, treating the mathematical programming equilibrium as if it were the emissions trading equilibrium may be a serious misrepresentation of the emissions trading process. A significant proportion of the gap between the expected cost savings and those achieved can be explained by the dynamics of the trading process. Specifically, we showed that a sequence of bilateral trades which conforms to current EPA regulations produces substantially less cost savings for a nonuniformly mixed pollutant than the cost-effective allocation. A sensitivity analysis which recalculated the results for a variety of assumptions about the target air quality levels and the nature of the trading process suggested that the results are robust, at least for our data set.

Theoretical and empirical considerations suggest three principal reasons why actual trades fail to capture the potential cost savings of theoretically modeled TDPs. First, cost-effective trades that could be consummated in a multilateral, simultaneous trading environment cannot be consummated in a bilateral, sequential trading environment because they would violate ambient air quality standards. Forcing each individual to trade to satisfy the air quality constraint, as is done by the current policy, is a much more restrictive condition than requiring that only the final equilibrium satisfies those constraints. Second, both current practice and our simulations require that emissions must either be held constant or reduced by each trade. Reaching the cost-effective allocation from the SIP allocation necessarily involves some trades that allow emissions to increase; these occur when the acquiring source is near air quality receptors that are already recording significantly better air quality than that required by the standards. Current practice rules out these cost-effective trades. Clearly, this constraint is more binding on nonuniformly mixed pollutants than uniformly mixed ones, since in the latter case the location of emissions does not matter. In fact, most recorded trades involve the latter type of pollutants. Third, all of the partial-information strategies require traders to secure all possible TDPs from their trading partner in one-time trades. Greater cost savings could result if some permits were reserved for subsequent trades.

1.5 Equity Concerns

Total compliance costs consist of all expenditures to control emissions and payments for permits on uncontrolled emissions. If TDPs are distributed without charge according to SIP uncontrolled emissions (that is, they are grandfathered), compliance costs only consist of control costs and firms will incur no additional costs relative to the SIP control strategy. However, if the government auctions off the permits, compliance costs consist of pollution control costs and payments for permits, and individual firm as well as regional compliance costs may rise. This is sure to elicit claims of inequity by adversely affected firms.

Tietenberg (1985) summarizes the results of a number of studies which compute the compliance costs of achieving a common air quality level using the SIP system relative to systems of APs and EPs which are auctioned to firms by the government. Typically, while AP systems reduce the financial burden, EP systems do not.

Atkinson and Tietenberg (1984) consider a stricter test of whether TDP systems that achieve the primary and secondary standards are Pareto-optimal relative to the *current* SIP level of control (where ambient air quality exceeds the primary standard). Three administrative rules are examined for improving air quality levels to the primary or secondary standards. The air quality degradation rule (AQDR) and the uncontrolled emissions rule (UER) are extensions of the SIP system, which comprises the third rule. The AQDR rule allocates individual source control in proportion to its contribution to total ambient degradation, while the UER rule allocates individual source control in proportion to its share in total emissions. The

UER rule scales this control responsibility until ambient standards are met. The EP system is modified to ensure that total emissions are the same as those achieved with the comparable SIP rule, but the AP system ensures only that the relevant ambient air quality standards are met and does not guarantee comparable emissions levels. The total number of permits is fixed by the air quality standard (in the case of APs) or the emissions target (in the case of EPs).

Based on the St. Louis data, our calculations indicate that if APs or EPs are grandfathered, near-Pareto optimality is achieved for the AQDR and UER allocations to reach the primary standard relative to the initial SIP control levels. Even if APs are auctioned, less than one-half of the firms are adversely affected under these two allocation systems for both TDP systems. Total regional compliance costs are reduced substantially under both methods of permit allocation, with marginally greater reductions under grandfathering. However, at the secondary standard, the required reduction in emissions is so large that it swamps the potential cost savings offered by both permit systems. Regional compliance costs and the number of adversely affected firms are dramatically increased. Under auctioning, the increase in regional compliance costs is 2–15 times greater than under grandfathering. Compared with the initial SIP control levels, improving air quality to the secondary standard will impose a large financial burden no matter which method of allocating control responsibility and permits is employed. However, from a comparison of the costs of SIP controls after they have been scaled up to achieve ambient standards, and the compliance costs of TDP systems at these same standards, TDP systems would appear to be far more equitable. These results are highly consistent with those obtained in Atkinson and Tietenberg (1982) for zonal permit systems.

1.6 Summary and Conclusions

Constraints on the AP and EP systems – reflecting limitations on local emissions, long-range acid deposition, and bilateral/simultaneous trading – can substantially erode the minimum-cost position of both systems to achieve ambient standards. If TDPs are auctioned off rather than grandfathered, the increase in compliance costs can easily threaten their political acceptability. Bearing this in mind, a number of conclusions and recommendations can be drawn:

1. The AP system is substantially more cost effective than the SIP and EP systems and is preferable as long as constraints on local increases in

emissions and long-range pollution deposition are not binding and the additional complexity of the AP system is not overly burdensome. The cost of the CAA's requirement that trades cannot increase emissions is seen to be substantial. Since there is no theorem guaranteeing that an EP system will be at least as cost effective as the SIP system, but there is such a theorem for the AP system, we should not eliminate the latter from serious consideration.

2. An administratively simple HAP system, which uses the most degraded air quality monitor, captured nearly all the cost savings of the full AP system, but removed substantially fewer emissions than the SIP system. As long as environmental expenses remain relatively small for firms, few are likely to relocate in response to a single binding constraint, creating a " hot spot" problem. The cost of constraining emissions to their SIP level is shown to be substantial.
3. The zonal EP systems we examined capture only a fraction of the potential cost savings and often do little to improve the hot spot problem.
4. For plants with tall stacks, our results indicate that constraints on long-run deposition substantially reduce the cost advantage of the AP system relative to those of the SIP system. This is consistent with the heavy environmental loading from tall stacks under the AP system without such constraints. Research involving the effects of an HAP system for achieving local standards and limiting long-range deposition seems to be called for.
5. The nondegradation offset trading rule captures a large portion of the cost savings of the AP system, without increasing emissions above the SIP levels. The cost of a requirement that each trade be subject to offset ratios of one or larger would most likely be substantial. The properties of a system combining the nondegradation offset trading rule with a single (or limited number of) long-range deposition permit(s) should be investigated.
6. Sequential trading, even with full information, appears to reduce substantially the cost savings of TDP systems. We should examine modified systems that move closer to simultaneous trading. For example, we could allow and even promote repeated, simultaneous trading, with the right to modify initial agreements during a specified short period. In an age of complex computerized stock and futures trading, we should not assume that an AP system, with its requirements that firms purchase permits at each binding receptor, is too complex to implement. Further, it should be remembered that APs are required only for binding receptors. We should attempt to facilitate trades by providing complete,

low-cost information on potential trades (i.e., minimal brokerage fees) and provide low-cost diffusion modeling to traders. Experimental economics laboratories would appear to be an ideal location for research in this area.

7. As ambient standards tighten from the primary to the secondary standard, and as compliance is more rigorously enforced, grandfathering of permits appears to be necessary to avoid complaints about inequity. Such permits must be equitably provided for new firms that enter an area.

References

Atkinson, S.E., 1983, Marketable Permits and Acid Rain Externalities, *Canadian Journal of Economics* **16**(4):704–722.

Atkinson, S.E., and Lewis, D.H., 1974, A Cost-Effectiveness Analysis of Alternative Air-Quality Control Strategies, *Journal of Environmental Economics and Management* **1**:237–250.

Atkinson, S.E., and Lewis, D.H., 1976, Determination and Implementation of Optimal Air Quality Standards, *Journal of Environmental Economics and Management* **3**:363–380.

Atkinson, S.E., and Tietenberg, T.H., 1982, The Empirical Properties of Two Classes of Designs for Transferable Discharge Permit Markets, *Journal of Environmental Economics and Management* **9**:101–121.

Atkinson, S.E., and Tietenberg, T.H., 1984, Alternative Approaches for Reaching Ambient Standards in Non-Attainment Areas: Equity and Efficiency Considerations, *Land Economics* **60**(2):148–159.

Atkinson, S.E., and Tietenberg, T.H., 1987, Economic Implications of Emissions Trading Rules for Local and Regional Pollutants, *Canadian Journal of Economics* **20**(2):370–386.

Atkinson, S.E., and Tietenberg, T.H., 1991, Market Failure in Incentive-Based Regulation: The Case of Emissions Trading, *Journal of Environmental Economics and Management* **21**:17–31.

Baumol, W.J., and Oates, W.E., 1971, The Use of Standards and Pricing for Protection of the Environment, *Swedish Journal of Economics* **7**:42–54.

Hahn, R.W., 1989, Economic Prescriptions for Environmental Problems: How the Patient Followed the Doctor's Orders, *Journal of Economic Perspectives* **3**:95–114.

Howe, C.W., 1993, Taxes Versus Tradable Discharge Permits: A Review in the Light of the U.S. and European Experience, Working Paper, University of Colorado, Boulder, CO, USA.

Krupnick, A.J., Oates, W.E., and van de Verg, E., 1983, On Marketable Air Pollution Permits: The Case for a System of Pollution Offsets, *Journal of Environmental Economics and Management* **10**:233–247.

McGartland, A., and Oates, W., 1985, Marketable Permits for the Prevention of Environmental Degradation, *Journal of Environmental Economics and Management* **12**:207–228.

Montgomery, D.W., 1972, Markets in Licenses and Efficient Pollution Control Programs, *Journal of Economic Theory* **5**:395–418.

Seskin, E.P., Anderson Jr., R.J., and Reid, R.O., 1983, An Empirical Analysis of Economic Strategies for Controlling Air Pollution, *Journal of Environmental Economics and Management* **10**:112–124.

Tietenberg, T.H., 1985, *Emissions Trading: An Exercise in Reforming Pollution Policy*, Resources for the Future, Washington, DC, USA.

US Congress, 1990, The Clean Air Act of 1990, Title IV, Acid Deposition Control, *Federal Register*, December, Washington, DC, USA.

Chapter 2

Economic Instruments for the Control of Acid Rain in the UK

Alberto Pototschnig
London Economics, London, UK

Abstract

UK environmental policy has traditionally focused on administrative controls. The paper considers the introduction of market-based instruments in the context of the control of SO_2 emissions from large combustion plants. These are responsible for more than 80% of total emissions in the UK. The analysis focuses on England and Wales, and considers three separate sectors: the electricity supply industry (ESI), refineries and other industry. It looks at the options for achieving the emission quotas for large combustion plants set by the Large Combustion Plant Directive. Electricity generators are already allowed some discretion, subject to agreement by the enforcing authorities, to allocate pollution control responsibilities among power stations within each company. Given the similarity in the power station mix between National Power and PowerGen, the advantages of extending the mechanism to include power stations of both companies would be limited. More substantial savings can be obtained when refineries and other large industrial plants, which have much higher abatement costs, are included in the scheme. An emission permit scheme which includes the ESI, refineries and other industry is therefore proposed, and is preferred to a mechanism based on emission charges. Some features of this scheme are then discussed.

Key words: acid rain, SO_2 emissions, economic instruments, tradable emission permits.

2.1 Introduction

Environmental objectives can be achieved through a variety of policy instruments: the two principal approaches are administrative controls and market-based instruments. UK environmental policy has traditionally focused on administrative controls. The widespread use of economic instruments would constitute a stark departure from the current stance.

Over the last few years, however, the use of economic instruments has begun to gain ground in environmental policy discussions. In 1990, the government recognized that market-based instruments "offer the prospect of a more efficient and flexible response to environmental issues" (Department of the Environment, 1990a, par. 1.29). More recently, the government indicated that it "believes that the time has come to deploy them more fully to achieve environmental objectives" (Department of the Environment, 1992b, par. 3.44).

This chapter considers these issues in the context of the control of acid rain in the UK. It focuses on sulfur dioxide, the main source of acid rain, and draws on the modeling work undertaken by the author and his colleagues at London Economics, for the Department of the Environment (1992a). The chapter is structured as follows. Section 2.2 reviews the policy context for the control of sulfur emissions in the UK. In Section 2.3, economic instruments for air pollution control are briefly discussed. The model used in the simulation is presented in Section 2.4, and the results in Section 2.5. In Section 2.6, practical aspects of a tradable permit system are discussed. Section 2.7 presents some concluding remarks.

2.2 Acid Rain in the UK: The Problem and the Policy Context

Acid rain arises from the deposition in various forms of pollutants such as sulfur dioxide (SO_2) and nitrogen oxides (NO_x), among which SO_2 constitutes the main source. SO_2 emissions in the UK fell by approximately one-third between the 1979 peak and 1989, from 5.5 to 3.7 million tonnes. Since 1989, the economic recession has produced further reductions, and by 1991 emissions totaled 3.6 million tonnes. Large static sources – power stations, refineries and other industrial plants with capacity in excess of 50 MW_{th} – are responsible for more than 83% of total emissions, a share that has increased over the years. It seems therefore appropriate here to concentrate on these sources. The policy context for these sources is represented by:

- The 1988 Large Combustion Plant Directive (LCPD), promulgated by the European Commission (EC, 1988) sets emission standards for new large combustion plants and national targets for emissions from existing ones. These targets are specified in terms of percentage reductions from 1980 levels, and are applied in the UK through the Programme and National Plan (Department of the Environment, 1990b). The Plan subdivided the emission targets for the UK into quotas for the electricity supply industry (ESI), refineries and other industries, and within the ESI for the two main generators, National Power (NP) and PowerGen (PG).
- The Environmental Protection Act 1990 (EPA) introduced the system of Integrated Pollution Control (IPC), based on authorizations and regulation. For air pollution, IPC replaces the regulations of the Health and Safety at Work Act 1974.

The standards for solid and liquid fuels for new plants, including those that came into existence after 1 July 1987, imply higher percentage reductions in emissions for larger plants than for smaller ones. This is a reflection of the economies of scale in sulfur removal, and indicates the fact that regulation has been planned with some regard to the cost of emission abatement. The sliding scale used in setting the standards does not appear, however, to have been designed specifically to achieve economic efficiency.[1] Somewhat more relaxed standards are set for new plants that burn indigenous solid fuels if, owing to the particular nature of the fuel, they are unable to meet the relevant emission levels "without using excessively expensive technology".

For existing plants, *Table 2.1* presents the target SO_2 emission quotas for England and Wales for 1991, 1993, 1998 and 2003 set by the National Plan, and compares them with actual emissions in 1991. Partly because of the economic recession in the UK, the National Plan quotas are currently higher than actual emissions.

For existing plants falling within the scope of the LCPD, the National Plan represents the UK government's current policy. In England and Wales, its implementation is the responsibility of Her Majesty's Inspectorate of Pollution (HMIP), which subdivides the sectoral ceilings into allocations to individual existing plants. This is done within the framework of the more general environmental policy set out in the EPA, which provides for large combustion plants to become subject to authorization under IPC. One of the main objectives of the EPA is to ensure that "the best available techniques not entailing excessive costs (BATNEEC) will be used for minimizing the pollution which may be caused to the environment taken as a whole by

Table 2.1. National Plan quotas and actual emissions in England and Wales (in 1,000 tonnes SO_2).

	National Plan quotas					Actual emissions,
	1991	1993	1995	1998	2003	1991[a]
NP	1,595	1,497	1,290	982	660	n.a.
PG	1,077	1,019	878	669	450	n.a.
ESI total	2,680	2,516	2,168	1,651	1,110	2,348
Refineries	86	86	84	82	78	64
Other industries	273	241	225	201	140	162
Total	3,039	2,843	2,477	1,934	1,328	2,574

[a]Provisional figures.
Source: Department of the Enviroment, 1990b; HMIP, 1993, Table 5.1.

the release [of prescribed substances], having regard to the best practicable environmental options available" [EPA, section 7(7)].

While there is still some uncertainty about the most appropriate approach to the interpretation of BATNEEC (see Pearce and Brisson, 1993), for our purposes it suffices to note that Process Guidance Notes, issued by the Chief Inspector, define the standards that BATNEEC is expected to achieve. Emission standards for new large combustion plants are the same as those contained in the LCPD. Existing plants (those in operation before 1 July 1987) are expected to achieve the standards for new plants as soon as the opportunity arises, and by 1 July 2001 at the latest.

Existing large combustion plants are therefore subject to the National Plan quotas. It is expected that they will also implement BATNEEC to achieve new plant standards by 2001 at the latest. Authorizations issued by HMIP should contain conditions to ensure that, among other things, each plant contributes to meeting the relevant National Plan quota. The UK has therefore chosen to implement international commitments by retaining an emphasis on administrative controls. The only element of flexibility so far is that each generator in the ESI can alter, during the year, the distribution of its overall quota among plants. In particular, subject to HMIP agreement, it is possible for a plant to "be operated after the [quota] allocation has been exceeded provided that [there are] equivalent offsetting decreases from other plants under the operator's control" (Department of the Environment, 1990b, par. 19). It is in this context that the potential role of incentive-based mechanisms has to be judged.

2.3 Economic Instruments and the Control of Acid Rain

Economic instruments for air pollution control may take various forms. On purely theoretical grounds, policies based on emission charges and policies based on tradable emission permits are recognized to lead to very similar results. As Weitzman (1974) noted, "generally speaking it is neither easier or harder to name the right price than to name the right quantities, because in principle the *same* information is needed to correctly specify either".

Since the information available to the regulator is usually imperfect, on practical grounds the choice is between uncertainty over costs and uncertainty over levels of emissions. The issue is whether it is more appropriate to accurately determine the level of costs that the industry would have to afford and accept some uncertainty on the outcome in terms of emissions, or to fix the maximum emission level and accept some uncertainty over the costs of achieving it. In this sense, permits deliver the prespecified levels of emissions, equal to the number of permits which have been issued, while charges are not able to achieve the same accuracy, unless the regulator can accurately estimate the abatement costs of all sources.

However, the quality of the ambient air, rather than emissions, is the crucial factor affecting human health, buildings and economic activities.[2] Air quality standards are set to prevent damage due to excessive pollution. For pollutants that are not uniformly mixed, the relationship between emissions of pollutants and the quality of the air at specific locations generally depends on the characteristics of the emitting sources, and the meteorological and geographical conditions in which these emissions take place.

Since economic instruments reallocate emissions among different sources, the resource cost savings that can be achieved by the implementation of incentive-based mechanisms should be judged against the effect on air quality, the deposition rate at different locations, and the damage caused. In environmental policy terms, the trade-off between cost savings and local air quality is a difficult one. Charges or permit systems may be based on emissions of pollutants from specific sources, or on the concentration in the air, or on the deposition of the same pollutants caused by these emissions.

To be completely effective, air quality control through economic instruments based on concentration or deposition levels would have to involve as many different charges or classes of permits as there are locations where air quality is monitored. Such systems are, however, very complex to implement and monitor. Furthermore, a large number of different spatial permit markets would not achieve the thickness required for liquidity. This means that economic instruments based on concentration or deposition levels have never been used in practice. Nor have attempts to introduce a spatial consideration in the trading of emission permits, by imposing constraints based on air quality standards on such trading, been successful.[3]

Strictly speaking, SO_2 is not a uniformly mixed pollutant. However, when emitted by high-stack sources, it spreads over large areas. The recent report of Environment for Europe (1993) quotes studies of the transport of acid pollutants which show that, on average, only 15%–20% of SO_2 and NO_x emissions remain in the local area, defined in terms of a 150 km × 150 km_2. Similarly, simulation results reported in Ramus (1991) indicate that less than 30% of the UK's SO_2 emissions are deposited on UK soil. It is therefore possible to treat SO_2 emitted from large combustion plants as a uniformly mixed pollutant, at least when designing policy frameworks within individual regions or countries. The simulations presented in this paper do not take the spatial dimension into account. Furthermore, in our modeling we restrict our attention to one type of economic instrument, namely emission charges. However, given the theoretical equivalence between policies based on charges and permits noted above, the results can also be interpreted in terms of tradable permits. The choice between charges and permits is considered in Section 2.6.

2.4 The Model

A simulation approach has been used to determine the impact of incentive-based policies to reduce SO_2 emissions. Based on the data briefly described below, the impacts of incentive-based mechanisms have been modeled by imposing a charge, denominated in £ per tonne of SO_2 emitted, on different sectors. The model and the data used in the simulation are described in detail in Department of the Environment (1992a).

In the UK, large combustion plants account for more than 83% of SO_2 emissions. Among these, the ESI is responsible for the largest share. The simulations focus on England and Wales, due to the lack of comparable data available for Scotland, and consider three separate sectors: the ESI, refineries, and other industries.

2.4.1 The electricity supply industry

The organization of the ESI in England and Wales is based on a two-tier structure, where generators sell electricity to regional electricity companies, generally through a "Pool" system. Electricity generation is in the hands of a small number of companies. At privatization, plants running on traditional technologies were assigned to two private-sector operators, NP and PG, while nuclear generation was left in the hands of the government and run by Nuclear Electric.

The Pool creates strict links between the actions of different generating units: subject to network constraints, generating units are dispatched to supply electricity in merit order, i.e. according to the price at which they offer to supply. Our modeling has been based on London Economics' electricity supply model (POOLMOD), a detailed simulator of the dispatch merit order of all the available units in the England and Wales electricity system. The model matches capacity to demand by using the lowest-cost sets available in each half-hour time period.

In a competitive market, the Pool arrangements would induce generating sets to bid at their marginal costs. However, there is some evidence from the short period since privatization that the two main generators have at times bid at prices somewhat higher than marginal costs. In the model, we have not taken into account the strategic considerations that may lead to this behavior, but rather assumed that the ESI is a competitive sector.

Our base for simulation has been the ESI in England and Wales as it is likely to be in the mid-1990s. The framework is completely static, so that no attempt has been made to model the change in plant mix over time, due to plant retirement and investment in new generating capacity. The analysis is structured as follows:

- Simulation of the base mid-1990s scenario in terms of output, merit order and SO_2 emissions, based on a plant-by-plant analysis of fuel inputs. The base scenario takes into account the restructuring of the ESI which is likely to occur after the expiry in 1993 of the (largely politically dominated) contracts with British Coal. This will leave the privatized generators free to switch to cheaper fuels, and to cheaper technologies.
- Simulation, on a plant-by-plant basis, of specific alternatives for emission reductions (see Section 2.4.3) and estimation of the associated costs taking into account the characteristics of the individual plants.
- Simulation, on a plant-by-plant basis, of the new generation profile once the abatement option has been adopted in response to the imposition

of an emission charge at a specific level. The choice of the emission reduction option has an effect on the position of the plant in the merit order and, consequently, on the frequency with which it will be called to run. This, in turn, affects the choice between abatement techniques (which may be characterized by both fixed and variable costs), and the payment of the charge, which varies proportionately with emissions and, for a fixed level of fuel sulfur content, with output.

- Generation of the new merit order and iteration to the final generator choice, and hence generation profile. We have assumed that individual plants seek to maximize profits. When faced with a charge on SO_2 emissions, each plant chooses the level of abatement that produces the maximum level of profit. This level depends not only on the level of the emission charge and the available abatement technologies and their costs, but also on the choices of the other plants operating within the Pool.

2.4.2 Refineries and other industries

The simulation of the impact of economic instruments on refineries and other industries has been based on an inventory of large combustion plants in 1990 made available by the Warren Spring Laboratory, Department of Trade and Industry. The inventory contains plant-specific data on capacity and fuel type for all boilers that fall under the LCPD.

Since we have assumed that refineries and other industry are price takers for their output, profit maximization is equivalent to cost minimization. These plants will therefore choose the most appropriate trade-off between emission reduction and the payment of the charge, in order to minimize costs for a certain given level of output.

2.4.3 Options for reducing SO_2 emissions

SO_2 emissions arise from oxidation during combustion of the sulfur contained in the fuel. Apart from the small proportion of sulfur retained in the ash, there is a direct relationship between fuel sulfur content and SO_2 emissions. Two ways to reduce SO_2 emissions from power stations, refineries and other industrial plants have been considered in the analysis:[4]

- By using low-sulfur coal in coal-fired boilers or low-sulfur oil in oil-fired boilers, thus reducing the amount of sulfur involved in the combustion process. With the wider adoption of environmental policies around the world, demand for low- sulfur fuels will increase, and a low-sulfur premium (LSP) will emerge in the market. In our simulations, we have assumed a LSP equal to 2% for each 0.1% reduction in the sulfur content.[5]

- By removing sulfur from the flue gases by installing flue gas desulfurization (FGD) equipment. This is the most widely used approach for controlling SO_2 emissions from large combustion plants. Costs for FGD technologies exhibit economies of scale with respect to both the sulfur content of the flue gases and the capacity of the plant. Least-cost emission abatement is therefore achieved by applying FGDs to large plants and to those burning high-sulfur fuels.

There are other emission reduction possibilities, such as switching to a different type of fuel with a lower sulfur content. This can be done either by substituting the fuel used in a specific plant or by replacing a retiring plant with a new one burning a different fuel. These options have not been considered in the analysis. The results of the analysis may therefore overestimate emission reduction costs.

2.4.4 Scenarios

As discussed in Section 2.2, emissions of SO_2 from existing large combustion plants are subject to the emission quotas set by the National Plan. These plants are also expected to implement BATNEEC standards not later than 1 July 2001.

Although the BATNEEC requirement greatly reduces the scope for any incentive-based instrument, by implying that each individual source reduces emissions as much as possible, it would be too optimistic to assume that all (or most of) the existing large combustion plants would have achieved new plant standards by the target date. In our simulations, therefore, we have taken the National Plan provisions as the background against which the potential role of policies based on economic instruments are analyzed.

Simulations have been conducted to recreate the following situations:

1. The current implementation of the National Plan, assigning separate emissions quotas to NP, PG, refineries and other large industrial sources. When applying for authorization under the IPC regime, NP and PG submit their choices about the way in which they intend to achieve their allocated quotas, whether by using low-sulfur fuels, by installing FGD equipment, or by rearranging the generation mix. It is also possible,

subject to HMIP's agreement, to reallocate emission quotas during the year between plants owned by the same generator.[6] The current regime can therefore be seen as a market mechanism of some sort, within each company's set of plants. In our simulations, therefore, each company is free within these quotas to decide how to allocate emissions to individual plants.

The National Plan quotas for NP and PG are reduced by 24% between 1995 and 1998 and by 49% between 1995 and 2003. We have therefore assumed that compliance with the National Plan implies that each company would have to reduce emissions, with respect to our base case, by 24% in 1998 and by 49% in 2003. The National Plan quotas for refineries and other industries imply percentage reductions with respect to our base case of 23% and 41%, respectively, in 1998 and by 27% and 59%, respectively, in 2003.

2. NP and PG are allowed to trade emissions within an ESI overall quota. Each company may choose to reduce emissions by more than the level assigned to it by the National Plan and to sell the excess reduction to the other company, which in turn will be able to reduce emissions by a correspondingly lower amount. In this case, the required reductions of 24% and 49% would have to be achieved by the ESI as a whole, but not by the individual companies. Emission reductions required from refineries and other industries remain unchanged.
3. A single England and Wales overall quota is implemented, within which the power companies, refineries and other industries included in the National Plan quotas are free to trade emissions.

Table 2.2 shows the level of (unabated) emissions from our base case scenario and compares it with the National Plan quotas for 1995. It also summarizes the percentage reductions from our base case which we assume would have to be achieved under the three scenarios considered.

The base scenario simulates the starting situation, in terms of emission levels, in which polluters make decisions about abatement actions, when faced by an emission charge. The assumed required percentage reductions are different from those indicated in the National Plan, since they relate to the base levels assumed in the simulations, and not to actual emissions in 1980. Furthermore, for the ESI, the assumed required emission targets have been derived by directly applying the percentage reductions in National Plan quotas between 1995, 1998 and 2003 to our base case emissions, even though the latter are not in line with the corresponding 1995 quotas.

Table 2.2. Base case emissions, National Plan quotas, and assumed required emission reductions in England and Wales.

	Emissions (1,000 tonnes SO_2)		Assumed required emission reductions[a] (%)					
	Base case emissions 1995	National Plan quotas 1995	1998 S1	1998 S2	1998 S3	2003 S1	2003 S2	2003 S3
NP	1,311	1,290	24	24 (NP and PG)	26 (all sectors)	49	49 (NP and PG)	49 (all sectors)
PG	653	878	24			49		
Refineries	107	84	22	22		27	27	
Other industries	338	225	41	41		59	59	

[a]S1 = separate quotas; S2 = unified ESI quota; S3 = unified England and Wales quota.

2.5 The Results

The aim is to compare the resource costs of achieving the required emission reductions under the different scenarios. These are the net costs of the measures aimed at reducing emissions; they exclude any additional costs arising from the impact of environmental policies on the operation of the power Pool system. These additional costs are highly sensitive to the nature of competition in the electricity market which, at present, is still an unresolved issue. An estimate of such additional costs in the case of a perfectly competitive ESI is included in the results presented in Department of the Environment (1992a).

2.5.1 Separate quotas

Given the difference in the characteristics of plants in different sectors, and in the mix of generating plants between NP and PG, the extent to which an emission charge would affect the level of emissions varies between the two power companies, refineries and other industries.

In the case of refineries and other industries, the cost minimization assumption implies that plants operate at points where the marginal cost of abatement is equal to the emission charge imposed. For the ESI, even though the more complex interrelationship between different plants means that profit maximization is not generally equivalent to cost minimization, the difference between marginal abatement costs and the emission charge level is minimal. The level of the emission charge is therefore a good approximation of the marginal cost of abatement.

The ESI is characterized by lower abatement costs than the other two sectors. In the ESI, there is a range of charge levels over which PG's (marginal) abatement costs are lower than those of NP. *Figure 2.1* illustrates the marginal abatement costs for NP, PG, refineries and other industrial plants and shows the charge levels required to comply with the separate National Plan quotas for 1998 (*Figure 2.1a*) and for 2003 (*Figure 2.1b*).

The National Plan assigns different emission reduction targets to the three sectors under consideration, in an attempt to take differences in abatement costs into account. Our results indicate that the National Plan has not gone far enough in this differentiation. In the case of industrial sources, it has gone in the opposite direction. It is true, however, that most of the emission reductions in the industrial sector are expected to come from technological progress, which is likely to be faster in industry than in refineries and the ESI. This aspect is not considered in the present analysis.

Marginal abatement costs corresponding to the required emission reduction levels are much higher for other industry and refineries than for the ESI. Within the ESI, costs are lower for PG than for NP. There is therefore a clear indication that pooling the quotas assigned to the power companies, to refineries and to other industry would improve economic efficiency. We now turn to these scenarios.

2.5.2 Unified ESI quota

NP and PG exhibit somewhat different abatement costs over the range of emission reduction levels that are relevant for compliance with the National Plan.

We have therefore investigated the possibility of allowing the two companies freedom in achieving the overall ESI quota. Each company decides its optimal level of emissions reduction with respect to the current charge level. Similarly, a permit market would give each of the two generators the possibility of exceeding or stopping short of its own quota and, respectively, buying or selling "emission credits" to or from the other company.

The National Plan quotas imply reductions in ESI SO_2 emissions by 24% and 49% by 1998 and 2003, respectively. At both of these reduction levels, PG is characterized by lower marginal costs than NP. Therefore, starting from a situation where both companies were required to achieve these reduction levels, the cost for PG of reducing SO_2 emissions by one extra tonne is lower than the reduction in abatement costs for NP from increasing emissions by a corresponding amount. In an emissions trading scheme, NP would therefore be willing to pay a price for acquiring the right to emit an extra tonne of SO_2 which is higher than the extra costs incurred by PG to offset NP's increased emissions by increasing abatement.

(a) 1998

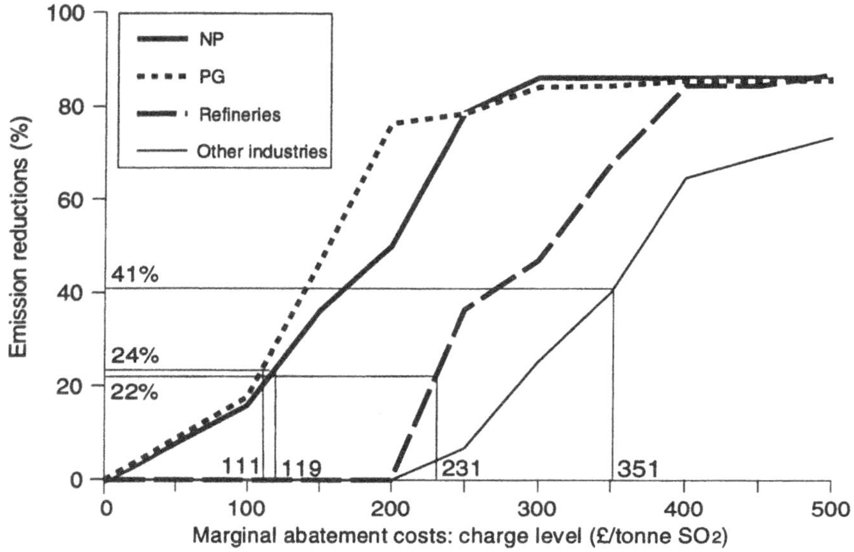

(b) 2003

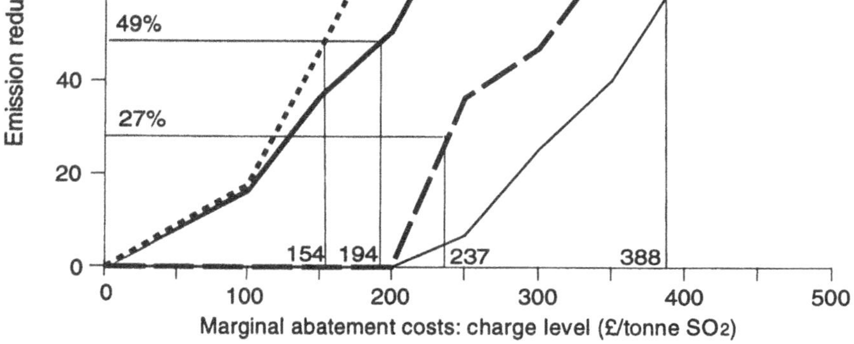

Figure 2.1. Marginal abatement costs and SO_2 emission reductions induced by different levels of charges: separate quotas (a) 1998, (b) 2003.

As the burden of abatement shifts in this way more and more from NP to PG, PG's marginal abatement costs would increase, while NP's abatement costs would decrease, until the point at which they are equal. This is the equilibrium situation in the scenario where a unified ESI quota is considered.

A charge of £115/tonne SO_2 applied to both NP and PG would deliver the 24% reduction in emissions from the ESI required to comply with the National Plan quotas for 1998. This would be achieved by:

- PG reducing SO_2 emissions by 26%, or by 18,000 tonnes more than the level required by its quota.
- NP reducing emissions by 22%, or by 18,000 tonnes less than the level required by its quota.

If this result were to be achieved through a tradable permit system, it would involve PG transferring the right to emit 18,000 tonnes to NP at the equilibrium price of £115/tonne SO_2 (equal to the equilibrium marginal abatement costs). The value of the transaction would therefore be in excess of £2.1 million.

Similarly, a charge of £174/tonne SO_2 would be required to achieve the National Plan ESI quota for 2003. Again:

- PG would abate SO_2 emissions by 60% or by 77,000 tonnes more than the level implied by its quota.
- NP would reduce emissions by 45%, or by 77,000 less than the level implied by its quota.

If this result were to be achieved through a tradable permit system, it would involve PG transferring the right to emit 77,000 tonnes to NP at the equilibrium price of £174/tonne SO_2. The value of the transfer would be in excess of £13.6 million.

Figure 2.2 compares emissions for NP, PG and the ESI as a whole with the level required by the National Plan quotas for 2003 and provides a graphical representation of the levels of charges that would be required to ensure compliance with these quotas.

2.5.3 Unified England and Wales quota

Refineries and other industries are characterized by much higher abatement costs (per tonne of SO_2) than the ESI, primarily due to the smaller scale of their plants.

In a unified England and Wales quota scenario, it is therefore economically efficient to shift some of the emission reduction responsibilities from refineries and other industry to the ESI.

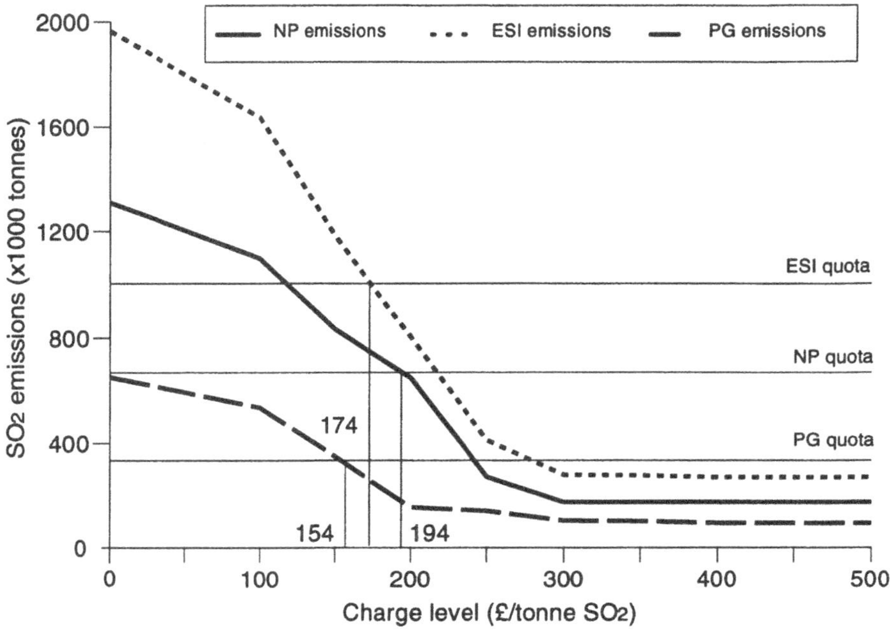

Figure 2.2. SO_2 emission reductions induced by different levels of charges and National Plan quotas: unified ESI quota, 2003.

A charge of £133/tonne SO_2 applied to the ESI, refineries and other industry would deliver the 26% reduction in emission levels in England and Wales required by the National Plan in 1998. This reduction would be achieved entirely by NP and PG:

- PG would reduce SO_2 emissions by 35%, or by 85,000 tonnes more than the level required by its quota.
- NP would reduce emissions by 29%, or by 76,000 tonnes more than the level required by its quota.
- The ESI would therefore emit 161,000 less than its quota.
- Refineries and other industries would not reduce emissions.

If this result were to be achieved through a tradable permit system, it would involve refineries and other industries buying "rights" to emit 24,000 and 137,000 tonnes, respectively, from the ESI, at the equilibrium price of £133/tonne SO_2. The value of the transfer would amount to £21.4 million.

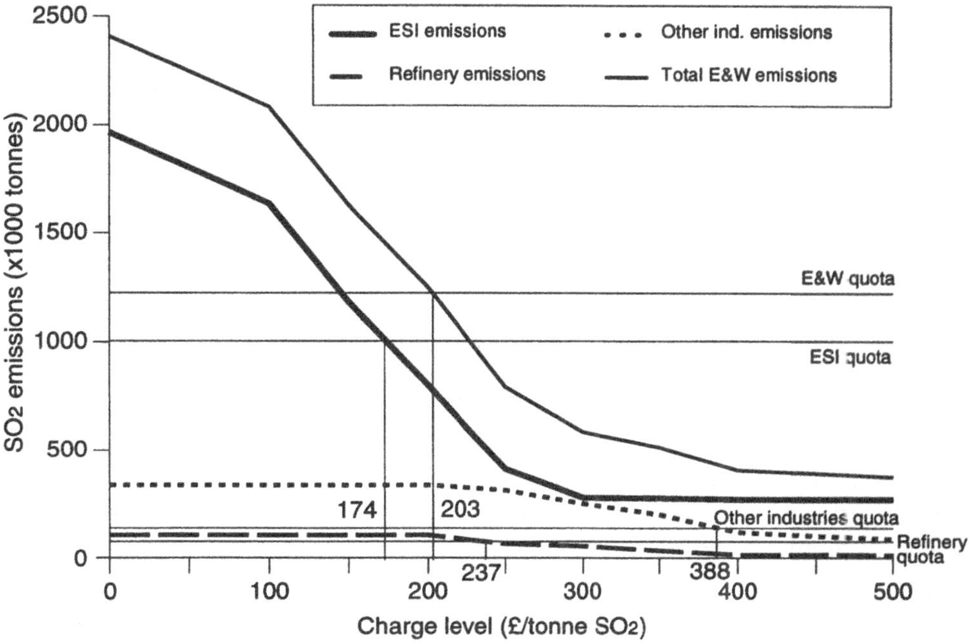

Figure 2.3. SO_2 emission reductions induced by different levels of charges and National Plan quotas: unified England and Wales quota, 2003.

For 2003, the required level of charge would be equal to £203/tonne SO_2, and a similar pattern of emission reductions and emissions trading would occur:

- PG would reduce SO_2 emissions by 76%, or by 181,000 tonnes more than the level required by its quota.
- NP would reduce emissions by 52%, or by 43,000 tonnes more than the level required by its quota.
- The ESI would therefore emit 224,000 less than its quota.
- Refineries and other industries would reduce emissions by 2% and 1%, respectively.

If this result were to be achieved through a tradable permit system, it would involve refineries and other industries buying "rights" to emit 27,000 and 197,000 tonnes, respectively, from the ESI at the equilibrium price of £203/tonne SO_2. The value of the transfer would amount to £45.5 million.

Figure 2.3 compares emissions for the ESI, refineries, other industries and for all sources with the level required by the National Plan quotas for 2003 and provides a graphical representation of the levels of charges required to comply with these quotas. The operation of a permit system under a unified quota is summarized in *Table 2.3*, where the pattern of emission trading is illustrated.

2.5.4 Resource costs

What are the resource cost savings arising from transferring the burden of emission reduction from one company to the other within the ESI, or from refineries and other industry to the ESI? *Table 2.4* presents the resource cost savings that a unified ESI quota would generate. The savings in resource costs associated with the introduction of a unified ESI quota are estimated to be around £0.1 million per annum (p.a.) for achieving emission reduction levels required by the 1998 quotas and £1.5 million p.a. for 2003. This is only 0.2% and 1.5% of the total resource costs for the ESI with separate quotas, estimated to be £32 million p.a. for 1998 and £103 million p.a. for 2003. It should be noted, however, that

- The existing quotas for NP and PG, and the interpretation we have given to them here, already allow for an optimal allocation of emission reduction responsibilities within each company.
- Marginal costs do not differ substantially between the two companies.

Larger savings can be achieved by widening the scope of emissions trading to refineries and other industry. *Table 2.5* presents the estimates for the resource cost savings in this case. Resource cost savings are now more substantial, due to the large differential in abatement costs between the ESI and other sources.

To comply with 1998 emission targets under an England and Wales unified quota, ESI emissions would be 161,000 tonnes lower than under separate quotas, and costs correspondingly higher by £20 million p.a. But refineries and other industry will face no need for emission reductions, and therefore will save £5 million and £39 million p.a., respectively. The net resource cost saving is therefore equal to £24 million p.a., or 32% of the total resource costs under separate quotas.

To achieve the emission reduction targets for the year 2003 under an England and Wales unified quota, ESI emissions would be 224,000 tonnes lower than in the case of separate quotas, and resource costs correspondingly higher by £42 million p.a. Refineries and other industry would save almost all costs associated with emission abatement, which in the case of

Table 2.3. Emission reductions and emissions trading in England and Wales (in 1,000 tonnes SO_2).

	Emission reductions, 1998			Emission reductions, 2003		
	Required under separate quotas	Achieved under unified quota	Difference (traded[a])	Required under separate quotas	Achieved under unified quota	Difference (traded[a])
NP	313	389	+76	640	683	+43
PG	155	240	+85	318	499	+181
Refineries	24	0	−24	29	2	−27
Other industries	137	0	−137	198	1	−197
Total	629	629	0	1,185	1,185	0

[a]+ (positive) signs denote net sellers; − (negative) signs denote net buyers.

Table 2.4. Resource cost savings from a unified ESI quota (£ million p.a.).

Emission reductions to satisfy quotas in:	1998	2003
Resource costs with separate quotas		
NP	21.6	71.7
PG	10.1	31.6
Total	31.7	103.3
Resource costs with unified ESI quota	31.6	101.8
Resource cost savings	0.1	1.5
(as % of costs under separate quotas)	0.2%	1.5%

Table 2.5. Resource cost savings from a unified England and Wales quota (£ million p.a.).

Emission reductions to satisfy quotas in:	1998	2003
Resource costs with separate quotas		
NP	21.6	71.7
PG	10.1	31.6
Refineries	5.4	6.4
Other industries	39.2	61.7
Total	76.3	171.4
Resource costs with unified England and Wales quota		
NP	31.4	80.3
PG	20.5	63.5
Refineries	0.0	0.5
Other industries	0.0	0.3
Total	51.9	144.6
Resource cost savings	24.4	26.8
(as % of costs under separate quotas)	32.0%	15.6%

the separate quotas would amount to £6 million and £62 million p.a., respectively. The total resource cost saving is therefore around £27 million p.a. This is 16% of the total resource costs involved in achieving emission standards, estimated at around £171 million p.a.

2.6 The Implementation of Economic Instruments

The results presented in the previous section suggest that a policy based on economic instruments extended to include the ESI, refineries and other large industrial sources may provide substantial resource cost savings in achieving the National Plan emission targets. The analysis was conducted using emission charges but the results can be interpreted in terms of emission permits as well. We now turn to the practical issues of environmental policy, and ask what the best options are for introducing economic instruments for air pollution control in the UK. In particular, which type of policy should be implemented, and what is the best way to implement the selected policy?

2.6.1 Emission charges versus tradable permits

The UK has committed itself under the LCPD to reducing SO_2 emissions according to a specific target. In this case, permits are more appropriate than charges, since they would deliver the required emission reduction levels.

Abatement costs vary over time because of technological innovation, exchange rate fluctuations and general inflation. Charges would have to be continuously adjusted to deliver a constant level of emissions. Furthermore, the UK's commitment to a staged reduction in emissions would require a frequent redefinition of the appropriate charge level. Permits avoid these problems; a specified emission profile over time can be achieved by simply varying the number of permits on the market.

Finally, permits may give a more flexible way of introducing market-based instruments without imposing immediate substantial financial costs on the polluters. This can be achieved by using appropriate allocation schemes.

2.6.2 The implementation of a tradable permit scheme

In designing an emission permit scheme, attention should be given to the denomination of permits and the effect of noncompliance; and the permit allocation method and the extent to which the permit market is open to manipulation.

Permit Denomination and Noncompliance

Permits may be defined as giving their holders the "right" to emit a specified quantity of pollutants in a specific year, or emissions over a period of more than one year. Multi-year permits are probably more appropriate, provided that they confer on the holder the "right" to a decreasing level of emissions over time. The decay rate built into the multi-year permits should be faster than the rate of reduction in emissions to which the UK is committed under international agreements, so that extra permits can be issued later for subsequent years.

The local nature of the effects of SO_2 may further complicate the operation of an effective tradable permit scheme. However, as noted in Section 2.3, SO_2 emissions from high-stack sources spread over large areas, traveling several hundred kilometers. As long as sources are located relatively close to each other, emission permit trading would have immaterial net effects on the level of deposition at any location.

Regional tradable permit schemes, for which the UK is divided into a small number of regions, should therefore be introduced. Permit trading could be allowed on a one-to-one basis within each region, while trading between regions would attract a penalizing exchange rate, which reflects the least favorable difference in the damage caused by emissions from sources in the different regions. This exchange rate will ensure that any trade would not increase damage at any location. The results in Section 2.5 indicate that the large savings in resource costs from a tradable permit scheme would come from the reallocation of pollution control responsibilities from refineries and other large industrial sources to the ESI. Most of the savings would therefore still be available if permit trading is restricted to the regional level, as long as each region has sources of all three types. This scheme should be preferred to one in which each individual source is assigned a score on the basis of the effect of its emissions on environmentally sensitive areas, and permits traded at rates of exchange that reflect the relative scores of the trading parties.

Whichever the permit scheme chosen, noncompliance may have adverse effects on the market's efficiency in allocating pollution control responsibilities. Noncompliant firms would internalize the possibility of being audited and fined. They would, in general, operate at a level where the marginal cost of abatement is not equal to the prevailing permit price. This would result in a suboptimal situation, in which marginal abatement costs are not equalized across sources. As Malik (1990) has shown, it is only when either the source's subjective probability of being audited is independent of its decisions, or when this probability is a function of the level of noncompliance, that the decision rule of the noncompliant source is the same as that of the compliant one. There would then be no effect on the efficiency of the solution attained by a tradable permit scheme. These considerations should be taken into account by the authorities in devising appropriate monitoring and enforcing methodologies. An auditing approach based on random selection of sources should be favored.

Permit Allocation and Market Manipulation

In competitive markets and with negligible transaction costs, the efficiency in the allocation of pollution control responsibility is independent of the way in which permits are allocated (see Coase, 1960). The largest savings in resource costs, however, would be achieved by a permit market that includes refineries and other industries as well as the ESI; in this market, the two electricity generators would account, at current emission levels, for approximately 85% of the total demand for permits. Even after the permit trading along the lines described in Section 2.5, the ESI would still hold 75% of the total number of permits in 1998 and 64% in 2003. It is therefore likely that NP and PG would have substantial power in the permit market.

Among the sectors that would participate in the permit market, the ESI is characterized by the lowest marginal abatement costs at current emission levels. If permits are initially allocated according to these levels, the ESI would act as a net seller of permits. The dominant position of the two generators may induce them to offer on the market a lower number of permits than if they were price-takers. This would result in a higher permit price, and an inefficient allocation of pollution control responsibilities, since marginal abatement costs would not be equalized across polluters (see Hahn, 1984). If, on the other hand, permits were auctioned off, dominant players would tend to buy fewer permits than if they were price-takers. The resulting permit price would be lower, but, even in this case, full efficiency would not be achieved. Dominant players in the permit market might also have a strategic interest in keeping the permit price at a high level. New entrants would have to buy a certain number of permits in the market, so a high price could discourage entry to the market.

In England and Wales, if permits were allocated on the basis of current emissions, both short-term considerations and the strategic considerations of the two electricity generators would lead to a permit price that is higher than in a perfectly competitive situation. Under an auction system, short-term and strategic considerations work in different directions, so that the negative effect on efficiency would be limited. As a general simple rule, therefore, tradable permits should be auctioned off. This would place all market participants on the same side of the market: it would increase the size of the market and reduce the incentives for dominant players to manipulate the permit market.

Different considerations may only apply to the initial transitional period. While permits have a value, it may take some time before experience in the new instrument and the way in which the market operates can provide reliable guidance on the level of permit prices. So this price, at least temporarily, might be characterized by wide fluctuations. Since the purchase of the appropriate amount of permits would constitute a substantial financial burden for the sources, it may be appropriate for the authorities to avoid the uncertainty in costs by issuing at least some of the initial allocation of permits to all sources at a fixed price. Trade would then take place at the emerging equilibrium price, but any volatility in financial costs and revenues would be limited to the marginal quantities bought and sold. Even in this case, the authorities may limit the extent to which dominant players may exert market power by implementing an initial allocation which gives the electricity generators, the dominant players in the permit market, a share of the total supply of permits which is substantially lower than that which would result from their current emissions.

2.7 Conclusions

While UK environmental policy has traditionally been based on administrative controls, the analysis presented in this chapter indicates that there are important resource cost savings to be achieved by moving toward an approach based more on economic instruments, and that includes the ESI, refineries, and other industries.

It is suggested that permits would be more appropriate than charges in controlling SO_2 emissions within the limits set by the LCPD and detailed in the National Plan. The results indicate that if the permit market is extended to include refineries and large plants in other industrial sectors as well the ESI, total abatement costs in England and Wales may be reduced by as much as 16%, or £27 million a year, by the year 2003.

Acknowledgments

This paper draws heavily on the modeling work undertaken at London Economics and partially sponsored by the UK Department of the Environment. London Economics' team included Zoltan Biro and Nicholas C. Owen and the author is greatly indebted to them for providing the intellectual stimulus for the development of such work. Useful comments on an earlier draft were provided by participants at the IIASA Conference on Economic Instruments for Air Pollution Control (Laxenburg, October 1993) and by two anonymous referees. The author is also grateful to Mary Barton for the artwork. Any views expressed in this paper are those of the author, as is the responsibility for any mistakes.

Notes

[1] HMIP (1991) recognized that, "due to its costs, it is normally more economical to fit [sulfur] abatement equipment to larger plant" (Annex 3, par. 2.3.3). Our FGD cost estimates, however, suggest that the average costs (per tonne of SO_2 removed) of achieving the required standards are still substantially higher for small plants than for larger ones. In particular, it costs, on average, £225 per tonne of SO_2 removed to reduce emissions from a 500 MW coal-fired boiler by 90%, while the average cost of reducing emissions from a 100 MW plant by 40% is around £600.

[2] It can in fact be argued that health is affected not only by ambient air quality, but more generally by total exposure to pollutants, which includes local effects such as indoor air quality. For a discussion of "exposure trading" see, for example, Routmasset and Smith (1990).

[3] The US experience of the emissions trading program under the Clean Air Act is relevant here. Trades in emission reduction credits were subject to stringent constraints on their effect on the level of air quality, and very few trades were executed. Hahn and Hester (1989) report that only two external offset transactions, involving a level of emissions less than 1% of the total, were made in 1985.

[4] An alternative approach is the implementation of new combustion technologies with inherently lower SO_2 emissions, such as fluidized bed combustion. Since these technologies are mainly applied to new plants, they have not been considered in the present analysis.

[5] Note that England and Wales have no indigenous source of low-sulfur coal or oil. The LSP on low-sulfur fuels which may be used in England and Wales will be determined by demand/supply conditions on the international markets from where such fuels will be obtained.

[6] It has been assumed, however, that NP and PG are committed under the National Plan to install FGD equipment at Drax, Ferrybridge, and Ratcliffe – with a total capacity of 7.6 GW – even though it is now doubtful whether PG will honor this commitment at Ferrybridge.

References

Coase, R., 1960, The Problem of Social Cost, *Journal of Law and Economics* **3**(1):1–44.

Department of the Environment, 1990a, *This Common Inheritance: Britain's Environmental Strategy*, HMSO, London, UK.

Department of the Environment, 1990b, *The United Kingdom's Programme and National Plan for Reducing Emissions of Sulphur Dioxide and Oxides of Nitrogen from Existing Large Combustion Plants*, Department of the Environment, London, UK.

Department of the Environment, 1992a, *The Potential Role of Market Mechanisms in the Control of Acid Rain in Britain*, HMSO, London, UK.

Department of the Environment, 1992b, *This Common Inheritance: Britain's Environmental Strategy, The Second Year Report*, HMSO, London, UK.

Environment for Europe, 1993, *Environmental Action Programme for Central and Eastern Europe*, Document submitted to the Ministerial Conference, 28–30 April 1993, Lucerne, Switzerland.

EC, 1988, Directive 88/609/EEC on the Limitation of Emissions of Certain Pollutants into the Air from Large Combustion Plants, *Official Journal of the European Communities* **31**(L336/1), Office for Official Publications of the European Communities, Luxembourg.

Hahn, R., 1984, Market Power and Transferable Property Rights, *Quarterly Journal of Economics* **99**:753–765.

Hahn, R., and Hester, G., 1989, Where Did All the Markets Go?: An Analysis of EPA's Emission Trading Program, *Yale Journal on Regulation* **6**:109–153.

HMIP, 1991, Process Guidance Note IPR 1/1: Combustion Processes, Large Boilers and Furnaces, 50MW(th) and over, HMSO, London, UK.

HMIP, 1993, *Annual Report 1991–92: National Review*, HMSO, London, UK.

Malik, A.S., 1990, Markets for Pollution Control when Firms are Noncompliant, *Journal of Environmental Economics and Management* **18**:97–106.

Pearce, D., and Brisson, I., 1993, *BATNEEC: The Economics of Technology-Based Environmental Standards*, CSERGE Working Paper 93-03, London and Norwich, UK.

Ramus, C.A., 1991, The Large Combustion Plant Directive: An Analysis of European Environmental Policy, Oxford Institute for Energy Studies, Oxford, UK.

Routmasset, J.A., and Smith, K.R., 1990, Exposure Trading: An Approach to More Efficient Air Pollution Control, *Journal of Environmental Economics and Management* **18**:276–291.

Weitzman, M.L., 1975, Price vs Quantities, *Review of Economic Studies* **128**:477–492.

Chapter 3

Modeling Air Pollution Abatement in a Market-Based Incentive Framework for the Los Angeles Basin

Sherman Robinson
IFPRI, Washington, DC, USA
Shankar Subramanian
Cornell University, Ithaca, New York, NY, USA
Jacqueline Geoghegan
University of California, Berkeley, CA, USA

Abstract

The chapter uses a regional computable general equilibrium (CGE) model of the Los Angeles basin to analyze the economic impact on the regional economy of instituting a "marketable permit" scheme to reduce air pollution. The model includes 16 producing sectors and considers three pollutants: nitrogen oxides, reactive organic gases, and sulfur dioxide. The model also includes abatement technologies in a nonlinear programming framework embedded in the general equilibrium simulation model. The key empirical finding is that for levels of abatement that are technically feasible, the total cost of abatement in the region is relatively small, of the order of a few hundred million dollars. With abatement, eliminating about 30% of emissions can be done with little reduction in regional product (1%–2%). After abatement possibilities are exhausted, additional reductions in emissions require much higher emission charges, leading to major changes in the structure of production in the region away from polluting sectors. The costs are also higher.

Key words: general equilibrium (CGE) model, Los Angeles basin, SCAQMD, RECLAIM, NAAQS, GAMS, nitrogen oxides, sulfur dioxide, reactive organic gases.

3.1 Introduction

This chapter describes a regional, environmental, computable general equilibrium (CGE) model for the Los Angeles basin in southern California designed to investigate the economic impacts and policy implications of instituting the proposed "marketable permit" scheme to reduce air pollution in Los Angeles. In Section 3.2, we discuss the severe air pollution problems in Los Angeles, the existing regulatory framework, and the newly proposed market-based program to bring the area into attainment for federally mandated levels of air quality. In Section 3.3, we briefly survey environmental CGE models and present our model. The full set of model equations are presented in an appendix. We then present some illustrative empirical results from our CGE model. We conclude by discussing the model's limitations and outlining steps for further research.

3.2 Air Pollution in Los Angeles

Los Angeles continues to be the most problematic air quality control region in the USA. In the 1990 federal Clean Air Act amendments, Los Angeles had the dubious distinction of being the only area in the USA with the designation of "extreme" for its level of ozone, and the area with the longest time available – 20 years – to attain the National Ambient Air Quality Standards (NAAQS) for ozone. These standards were originally set after the 1970 Clean Air Act amendments, and Los Angeles continues to have difficulty in attaining them. The original response to these standards was to set up a command-and-control (CAC) strategy, where the technology for pollution abatement was specified by the regulatory agency. This approach did succeed in reducing pollution somewhat, but at great cost, and the standards still were not all met. There has been growing interest in using a market-based approach, which economists have long advocated as a more efficient way to achieve target levels of pollution abatement. For surveys, see Tietenberg (1985) and Hahn and Hester (1989). Montgomery (1972) provides a theoretical treatment.

On 1 January 1994, the South Coast Air Quality Management District (SCAQMD) implemented a market-based system using emission permits in order to meet air quality standards in an economically efficient manner. This system regulates the precursors to ozone: nitrogen oxides (NO_x), reactive organic gases (ROG); and sulfur dioxide (SO_x). The initial market includes only NO_x and SO_x; ROG will be considered later. Ozone is not directly emitted by sources; rather, it is created by chemical reactions among these precursors in the presence of sunlight. The new approach to air pollution reduction in the Los Angeles basin, RECLAIM (Regional Clean Air Incentives Market), is designed to give firms flexibility in attaining the region's air quality goals. Firms can choose to add more stringent controls, reformulate products, and/or acquire emission reduction credits (ERCs) to achieve these goals. Firms willing to reduce their emissions below the level required by the district will earn ERCs, which they can sell to other firms that find it relatively more expensive to reduce emissions internally. Under this system, economists argue that the target level of emission reduction will be reached at a lower total cost than under command-and-control regulation.

The firms that are required to be in the new program are those stationary sources that currently have district permits for NO_x and SO_x and also have annual emissions of greater than 2–4 tons per year (depending on the sector). For NO_x, these include 390 facilities covering 64% of permitted emissions; and for SO_x, 41 facilities covering 83% of permitted emissions, which are mostly refineries, electric utilities, and chemical plants. Each regulated facility will have an initial allocation of emissions and its own rate of reduction. The annual rate of reduction for 1994–2003 is 8.3% for NO_x and 6.8% for SO_x. For NO_x, the state and federal standards are expected to be met by 1995. The federal NAAQS and the state standard for ozone are not expected to be met until 2010. The area is expected to come into compliance for SO_x by 2005. The district has estimated that RECLAIM will cost 42% less than the command-and-control strategy now in place (SCAQMD, 1993).

3.3 The LA-CGE Model

We begin with a review of existing environmental CGE models, all of which are economy-wide models. We then present our regional environmental CGE model: Section 3.3.1 first provides a short description of the regional CGE model, and Section 3.3.2 a more detailed description of the pollution abatement model.

3.3.1 Environmental CGE models

In a seminal article, Leontief (1970) presented a multisector input–output model that incorporated environmental externalities. There was an active literature in the 1970s using input–output models to analyze pollution. This work led to the development of economy-wide, environmental CGE models.

The literature on multisector environmental models is briefly surveyed by Robinson (1990), who formulates a small CGE model based on the Leontief model to explore issues of optimal abatement policy using pollution taxes. Some of these models focus on analyzing the impact of the Clean Air Act (e.g., Hollenbeck, 1979; Hazilla and Kopp, 1990). Jorgenson and Wilcoxen (1990a,b) developed a dynamic CGE model of the USA to explore the costs to the US economy of environmental regulations.

Pollution control regulations influence not only firms that are directly affected by the regulations, but also other, perhaps nonpolluting, firms that do business with the regulated industries. Aggregate pollution generation is obviously strongly influenced by changes in the sectoral structure of production, which, in turn, depend on changes in the structure of demand for final and intermediate goods. Changes in production structure will also be affected by the workings of factor markets. By including multisectoral goods and factor market linkages, a CGE model captures both the direct and indirect effects of changes in government policy. These general equilibrium linkages work through changes in prices, both of goods and factors of production, upstream of and downstream from the regulated sectors. Such price changes in linked markets are potentially important and difficult, if not impossible, to analyze in a partial equilibrium framework.

CGE models were first formulated for national economies. Dervis *et al.* (1982) provide a textbook description of CGE models. A CGE model simulates a market economy where prices and quantities adjust to clear markets for goods and factors of production. It includes consumers, whose decisions determine the demand for final goods, and profit-maximizing producers, whose decisions determine the supply of goods and the demand for intermediate and primary inputs. The government appears explicitly, generating revenue through various taxes, purchasing goods, and saving (or dissaving). The rest of the world is treated as a supplier of imports and a buyer of exports. A CGE model is complete in that it captures all transactions in the circular flow of income. The model used here is Walrasian in that it determines only relative prices, and the absolute price level is set exogenously.

The LA-CGE model is closest in spirit to a series of environmental economy-wide models developed by Lars Bergman (1988, 1990, 1991), who took an energy model and adapted it to include air pollution. Bergman (1990) estimates the impact on Sweden of achieving an 80% reduction in SO_x and a 30% reduction in NO_x emissions between 1980 and 1993, while keeping CO_2 emissions at their 1988 levels. In his model, he simulates the operation of an emissions permit market. He specifies an initial supply of permits that is equal to the total amount of permitted pollution and then solves for their price in equilibrium, under various assumptions about abatement costs and the tightness of the constraints.

3.3.2 The Los Angeles regional CGE model

Our regional CGE model includes the counties of the South Coast Air Quality Management District (SCAQMD, consisting of Los Angeles, Orange, Riverside, and San Bernadino counties). Many CGE models focus on issues of international trade (for a survey of trade issues in CGE models, see Robinson, 1989); the LA-CGE model adapts a standard trade model to the requirements of a regional model. Locally produced commodities are assumed to be imperfect substitutes for imported commodities, whether imported from the rest of the USA or from foreign countries. Similarly, goods produced in LA and exported are distinguished from goods produced and sold in the region. The responsiveness of trade ratios to changes in the ratio of local and external prices is determined by sectoral substitution elasticities. Sectors with low elasticities and low trade shares (e.g., the service sector) are relatively sheltered from the external market. The local price for sectors with high trade shares and/or high substitution elasticities will be largely determined by the external price, with exports and/or imports varying to clear the local market.

Our regional LA-CGE model has a number of special features, given that the region is embedded in a national economy. For example, unlike a national model, Los Angeles is assumed not to have its own currency, so its exchange rate in the model is fixed with respect to that of the USA. In addition, the region's aggregate producer price index is exogenous and fixed at unity. Therefore, the region's balance of trade is then determined endogenously. On the trade side, we assume high substitution elasticities for the manufacturing sectors, but also specify some imports as completely noncompetitive; that is, they are not produced in the LA region.

Two alternative specifications can be used for the labor market. In one, wages are exogenous and migration is assumed to clear the regional labor market. This may be a more realistic approach than assuming flexible market-clearing wages for some categories of labor which are highly mobile, such as highly skilled workers and professionals. The low-skilled and poor are probably less mobile in the short to medium run. In any case, the empirical results indicate little change in the aggregate real wage, so either fixing the real wage rate or the employment level has very little impact. However, the model in its current form does not disaggregate labor by skill category, which might have a greater impact on the wage rate outcome. The results presented here use the full employment variant of the model.

Aggregate investment in a small regional economy is not necessarily determined by aggregate regional savings. Aggregate investment is exogenous in the model, with capital flows from the rest of the country ("foreign savings") balancing savings and investment in the LA region. The model is run in comparative statics mode, assuming fixed sectoral capital stocks that are not affected by investment in the current period.

3.3.3 The pollution abatement model

We consider three pollutants, NO_x, SO_x and ROG. The objective is to reduce aggregate emissions of each pollutant at least cost to meet exogenously specified emission targets. It was shown by Montgomery (1972) that this objective can be met through a system of tradable emission permits when there is perfect information on costs. Baumol and Oates (1971) consider the case with imperfect information and show that an iterative procedure can be used to determine an emission charge that would have the same effect as allowing a market in emission permits. In a static framework with perfect information, a system of emission charges is equivalent to a system of emission permits. The only difference lies in the wealth effects, which would differ depending on the initial allocation of permits and the redistribution scheme according to which the emission charge revenue is returned to polluters. In a dynamic context, other differences between emission charges and permits are in how they affect firms' incentives to search for better abatement technology and in how oligopolistic firms respond under these schemes (see Milliman and Prince, 1989; Malik, 1990).

Pollutant emissions are modeled as follows. Pollutants are produced by "control categories," which are usually sources or processes, such as a cement kiln (a source) or surface coating of wood furniture (a process). In a given sector, several control categories produce any particular pollutant. It is assumed that each control category's share in total emissions before abatement for a given sector and pollutant is constant at its value in the base year (1989). In addition, total emissions in any given sector are strictly proportional to output in that sector. For simplicity, consider the single-pollutant case. For each control category (denoted by k), at most three abatement technologies (denoted by t) are available; these are referred to as the least stringent, mid-level, and most stringent technologies. Abatement follows a Leontief technology. For each control category and technology pair (k, t), the maximum possible reduction in emissions is a fraction α_{kt} of the initial emissions. In this range, the marginal cost of abatement is constant and is equal to the unit cost of the inputs required in the abatement process. Notice that this marginal cost is the same across all sectors.

Since the marginal cost of abatement is constant for a given control category and technology, a firm facing an emission charge will abate to the fullest possible extent if the emission charge exceeds the marginal cost of abatement. Correspondingly, any technology with marginal cost exceeding the emission charge will not be used by the firm. Let X_i be output and Q_i be emissions in sector i. There is a fixed emission coefficient, β_i, so that:

$$Q_i = \beta_i X_i. \tag{3.1}$$

Of these emissions, a fraction θ_{ik} is from control category k. The maximum possible level of abatement of emissions from this control category in this sector is $\alpha_{kt}\theta_{ik}Q_i$. Abatement A_{ikt} is either zero or takes this maximum value. This is expressed using binary variables y_{kt}; $y_{kt} = 1$ if emissions in control category k are abated using technology t, and $y_{kt} = 0$ otherwise. We have:

$$A_{ikt} = y_{kt} Q_i q_{ik} \alpha_{kt}. \tag{3.2}$$

Total abatement in sector i is simply the sum of abatement over all control categories and technologies. The total emissions after abatement are:

$$\overline{Q} = \sum_i \left(Q_i - \sum_{k,t} A_{ikt} \right). \tag{3.3}$$

In addition, there is one other constraint. It is possible that two different abatement technologies for a particular control category may be mutually exclusive; i.e., the reduction achieved from the more stringent technology cannot be added to that achieved by the less stringent one. For such control categories, only one of A_{k1}, A_{k2}, or A_{k3} can be nonzero. This can be readily expressed in terms of the binary variables y_{kt} whenever the technologies are mutually exclusive:

$$y_{k1} + y_{k2} + y_{k3} \leq 1. \tag{3.4}$$

Since abatement follows a Leontief technology, the marginal cost of abatement P_{kt} is the value of inputs required for one unit of abatement:

$$P_{kt} = \sum_i \rho_{kti} p_i. \tag{3.5}$$

Here ρ_{kti} is the demand generated in sector i by a unit increase in abatement for control category k and technology t, and p_i is the price of good i.

A firm facing an emission charge P_a per unit of pollutant emitted will use a particular abatement technology only if its marginal cost does not exceed the emission charge:

$$y_{kt} = 0 \text{ if } P_{kt} > P_a, \tag{3.6}$$

$$y_{kt} = 1 \text{ if } P_{kt} \leq P_a. \tag{3.7}$$

The inequality in equation (3.7) must be modified when the technologies are mutually exclusive. In this case, if $P_{kt} \leq P_a$, only one of y_{k1}, y_{k2}, or y_{k3} can be positive. This case is not considered further here because it appears that for all control categories in the pollutants that we are considering, the technologies are additive and not mutually exclusive.

Because sectoral output and pollution emissions are linearly related (equation 3.1), the emission charge of P_a per unit of pollutant translates into a tax τ_i per unit of output given by:

$$\tau_i = \beta_i \left(1 - \sum_{k,t} y_{kt}\theta_{ik}\alpha_{kt} \right) P_a. \tag{3.8}$$

The cost of abatement in sector i is:

$$C_i = \sum_{k,t} P_{kt} A_{ikt}. \tag{3.9}$$

Pollution-intensive sectors thus pay a higher tax on their output, but this is modified by a factor that accounts for how effective abatement is in reducing emissions in that sector. The tax is collected by the government, which then proceeds to hand it back (net of abatement costs) as a lump-sum transfer to firms. The outcome is identical to that under a system of marketable permits for a particular initial distribution of emissions permits, which determines the lump-sum transfer shares.

The emission charge has two effects. At the sectoral level, it induces firms to choose an appropriate abatement technology. The emission charge also translates into a sector-specific per-unit tax on sectoral output. When this tax is at a low level, structural change is minimal and almost the entire reduction in emissions is obtained through abatement. However, abatement possibilities are soon exhausted, as will be seen below, and further reductions in emissions can be obtained only through relatively large structural changes induced by large changes in relative prices between sectors.

3.3.4 Implementation

The LA-CGE mode is implemented using the GAMS (general algebraic modeling system) modeling language (Brooke *et al.*, 1988). The abatement model is an activity analysis optimizing model, which has to be integrated into the CGE market simulation model. The difficulty is that complementary slackness conditions such as equations (3.6) and (3.7), which really involve dual variables, cannot be included directly in the primal of a programming model. However, it is easy to show that the following nonlinear programming problem will generate a solution that satisfies the inequalities in equations (3.6) and (3.7):

$$\max \sum_{i,k,t} A_{ikt}$$

$$\text{s.t.} \quad P_{kt}y_{kt} \leq P_a y_{kt} \qquad \text{and} \qquad P_{kt}(1 - y_{kt}) \geq P_a(1 - y_{kt})$$

The maximand is the sum of abatement over all sectors, control categories and technologies. The condition that $P_{kt}y_{kt} \leq P_a y_{kt}$ leads precisely to the inequality in equation (3.6) while the condition that $P_{kt}(1 - y_{kt}) \leq P_a(1 - y_{kt})$ leads to the inequality in equation (3.7). Generalizing to the many-pollutant case is straightforward because there is no interaction between abatement for different pollutants and the objective function is additive. The inequalities above are now specified for each pollutant and the objective function becomes the sum of abatement over all pollutants, sectors, control categories, and technologies. The per-unit tax on output is now obtained by summing up the tax from equation (3.8) over all pollutants. Once the emission charge for each pollutant is specified, the model solves for the resulting abatement.

While this optimization problem (including the constraints of the CGE model) is a mixed binary (integer) nonlinear programming problem, one can ignore the binary nature of the y_{kt} and treat them as continuous variables ranging between 0 and 1. Nonbinary values for the y_{kt} can occur only when the abatement cost and emission charge coincide. Although abatement costs for many control categories are identical in the data for the base year, away from the initial equilibrium these ties are broken because sectoral prices change. No more than one or two of the y_{kt} are found to be nonbinary, which is a small price to pay for the reduction in solution time resulting from ignoring the binary nature of these variables.

In addition, if target levels are specified for the total emissions of each pollutant, minimizing a suitable objective function subject to inequalities (3.5) and (3.6) will give the emission charges required to meet those emission

Table 3.1. Sectoral composition of emissions and value added and sectoral import shares (in %).

		Sectoral composition				
Sector	Sector name	NO_x	ROG	SO_x	Value added	Import shares
AGFD	Agriculture and food	1.4	0.9	1.4	1.9	68.7
OILGAS	Mining, oil, and gas	8.7	7.3	6.1	1.1	97.8
LMANUF	Light manufactures	3.1	6.6	0.5	3.0	58.5
WOOD	Wood and furniture	0.3	11.1	0.0	0.8	56.0
CHEM	Chemicals	1.8	6.1	5.7	1.8	66.3
PETR	Petroleum products	24.1	12.2	39.8	1.1	9.7
GLASS	Glass and cement	9.4	1.1	6.6	0.5	79.8
SVCS	Other services	5.1	2.9	1.3	50.9	3.9
UTIL	Utilities	27.6	3.1	32.2	1.4	23.4
PDUR	Producer durables	6.1	25.4	1.3	12.6	67.9
CDUR	Consumer durables	0.3	2.2	0.0	0.4	84.1
TRADE	Trade	1.0	4.7	0.6	13.5	2.8
PERS	Personal services	0.2	2.6	0.2	0.6	7.9
REPAIR	Automobile repair	0.2	4.6	0.0	1.5	0.0
AIRTR	Air transportation	10.8	9.0	4.1	0.6	7.1
PUBADM	Public administration	0.0	0.0	0.0	8.3	0.0

Note: Import shares are shares of total supply coming from outside the Los Angeles region. Sectoral share columns sum to one except for rounding.

targets. Suitable objective functions are total emission charge revenue, total cost of abatement, and (the negative of) real regional GDP. In the single-target case, one can also minimize the emission charge itself. In general, the solutions from these different objective functions will not coincide when targets are specified for all three pollutants because the three targets may not be binding constraints at the same time.

3.3.5 The data

The base year for the economic data is 1990, which is the latest year for which the database developed by the US Forest Service (called IMPLAN) provides input–output data at the county and state levels. The IMPLAN database contains a 528-sector input–output table, including demand matrices for competitive imports and a vector of total noncompetitive imports by use.[1] These were aggregated to 16 sectors (shown in *Table 3.1*).[2] The prime consideration in the choice of this aggregation was to identify sectors which are major producers of NO_x, SO_x, and ROG.

Table 3.2. Maximum abatement by sector (% of initial emissions).

	NO_x Technology			ROG Technology			SO_x Technology		
Sector	1	2	3	1	2	3	1	2	3
AGFD	13.5	22.2	27.2	24.8	25.9	26.1	25.5	33.2	33.2
OILGAS	19.0	30.0	37.8	1.4	1.4	1.4	19.5	26.4	26.4
LMANUF	16.9	26.7	33.4	24.1	32.0	46.1	38.9	46.6	46.6
WOOD	29.4	33.3	35.2	16.5	18.2	18.2	75.0	75.8	75.8
CHEM	10.6	14.3	16.5	17.2	18.5	20.0	0.6	0.7	0.7
PETR	67.3	71.4	73.4	0.9	0.9	0.9	50.4	63.9	69.1
GLASS	66.5	70.7	72.7	25.2	26.1	26.9	74.2	77.1	77.1
SVCS	24.0	32.7	37.3	3.3	3.3	3.3	33.2	36.5	36.5
UTIL	32.0	42.9	45.9	0.0	0.0	0.0	4.2	4.6	4.6
PDUR	26.2	31.2	33.9	40.7	43.9	44.0	37.7	43.5	43.5
CDUR	28.8	32.1	33.9	20.1	36.4	94.4	21.2	22.9	22.9
TRADE	51.3	57.1	57.9	17.5	21.0	24.0	49.3	52.5	52.5
PERS	22.2	29.7	35.2	0.6	0.6	0.6	12.3	14.9	14.9
REPAIR	54.8	60.3	60.7	3.0	17.6	27.6	85.7	86.7	86.7
AIRTR	9.4	9.6	9.7	0.3	0.3	0.3	10.7	14.2	14.2

Note: Full sector names are given in *Table 3.1*.

Data on emissions of these pollutants for 65 industry groups and on abatement costs and input requirements for 87 control categories for 1989 were provided by the California Energy Commission (CEC). These data were also aggregated to conform to the above sectoral specification. Thus the model mixes production structure and technology data for 1990 with emissions and abatement data for 1989.[3]

Table 3.2 shows the maximum abatement possible (as a fraction of initial emissions) for each sector and pollutant and for each level of abatement technology. Abatement possibilities are rather limited for many highly polluting sectors. The petroleum sector, for example, produces 12% of total ROG, but only 0.9% abatement is possible. Similarly, the utility sector produces 32% of SO_x emissions, but abatement can reduce SO_x production in this sector by only 4.6%. SO_x production in the petroleum sector, on the other hand, can be cut in half using just the least stringent technology. Another major difference between sectors is in their import shares. The impact of an emission charge will be stronger on a sector with a high import share (holding pollution intensity constant). Such sectors will contract substantially because domestic production can be readily replaced by imports.

3.4 Results

3.4.1 The no-abatement model

We ran a series of experiments assuming no abatement possibilities. The no-abatement model solves in a similar manner to the abatement model, except that the abatement related equations are dropped, all the abatement variables are set to zero, and the optimization problem is changed to one of minimizing the total emission charge. In this model, the only way to reduce aggregate emissions of any pollutant is to change the sectoral structure of production in the LA region away from pollution-intensive sectors. While this treatment is an extreme simplification, it is useful because the model then provides an upper bound to the estimated sectoral adjustments required to reach air quality goals. In addition, when the proposed reduction are so stringent that abatement possibilities are exhausted, the behavior of the model reverts to that of the no-abatement model.

In the set of experiments performed under this model, a single emission charge is imposed on each of the three pollutants separately. The model solves for the amounts of emissions of all pollutants and for the total emission charge revenue from taxing the single pollutant. For each pollutant, we ran a series of experiments with increasing pollution charges, terminating with a charge that yields a 25% reduction in the generation of that pollutant. The results are presented in *Figure 3.1.* The charges required to achieve a 25% reduction are: $250,000 per ton for NO_x, $970,000 per ton for SO_x, and $360,000 per ton for ROG.

Figure 3.1 also shows the amounts of other pollutants generated under each charging scheme. The results show the strong complementarities between NO_x and SO_x in production. The percent of baseline emissions produced under each of the charging schemes for NO_x and SO_x run almost parallel, indicating that the two pollutants are produced together in the same sectors.

The results shown in *Figure 3.1* indicate the outer bound of emission charges required to achieve pollution reduction, given that there is no possibility of abatement. The reduction is achieved solely by changing the structure of demand and production. In a model with abatement, the total cost figures will depend on abatement costs and will be much lower. However, the results presented here remain relevant, since they indicate the charges required to achieve further reduction in pollution after abatement possibilities are exhausted.

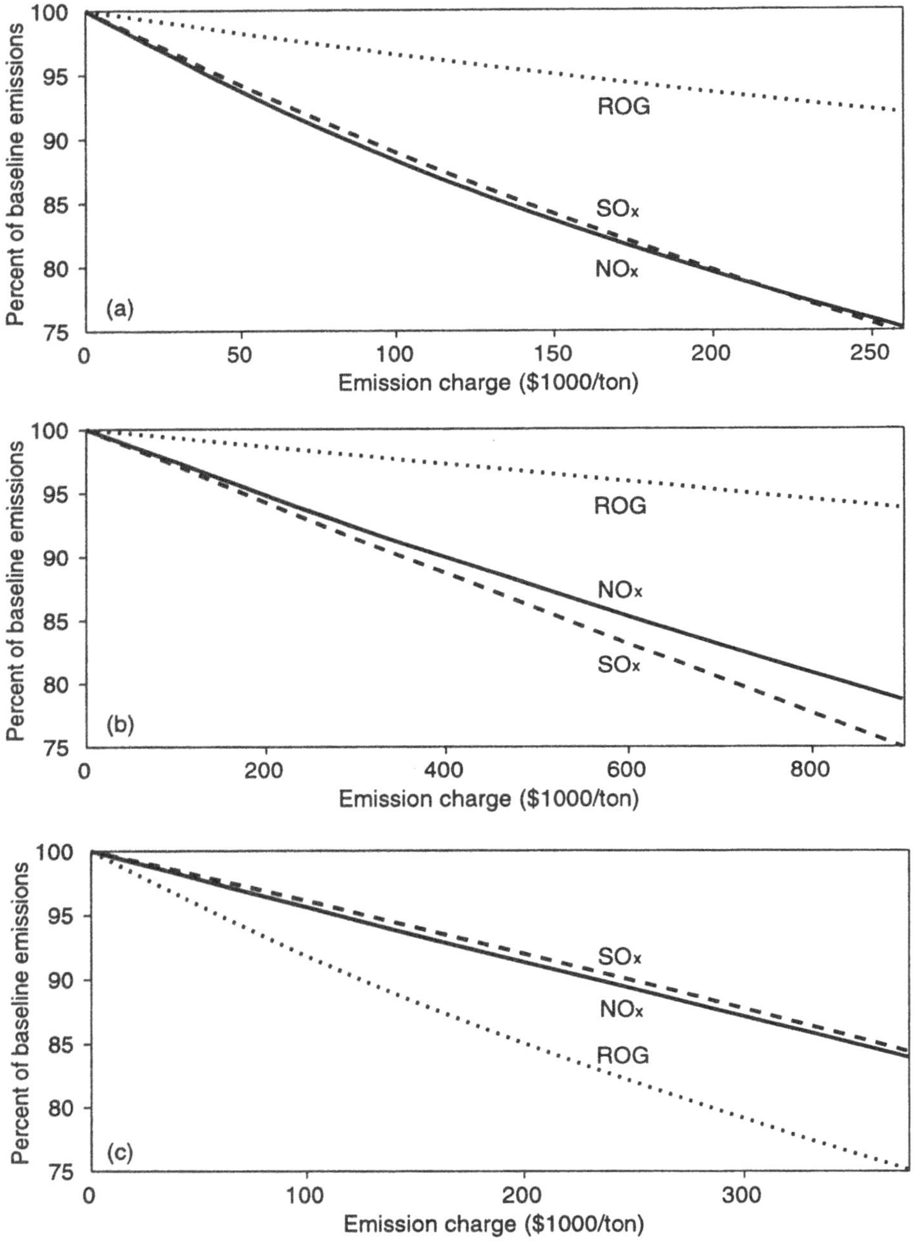

Figure 3.1. Emission charges and emission reductions. (a) NO_x, (b) SO_x, (c) ROG.

3.4.2 The abatement model

(a) NO_x Emission Charge

Figure 3.2a shows the emission charge required for a given level of abatement for reductions in pollution ranging up to 60%. Using the least stringent technology, the emission charge increases smoothly to about $17,000 per ton for a reduction of 40%. For reductions greater than 40%, the emission charge increases sharply, shooting up to $100,000 per ton for a 45% reduction. The reason for this is that up to this point, technologies for abatement are available. Any further reduction in emissions can be brought about only by changing the production structure; i.e., by decreasing output in NO_x-intensive sectors and increasing output in other sectors, and this calls for a substantially higher emission charge. Introducing the middle-level technology has the effect of pushing out this limit to about 45% reduction. Finally, making available the most stringent technology flattens the curve further, up to a reduction of about 50%.

Turning to the cost of abatement (*Figure 3.2b*), the curves for the three levels of technology largely coincide up to an emission charge of $10,000 per ton. This occurs because many of the mid-level and most stringent technologies have abatement costs above this level, and are therefore not used until the emission charge goes beyond this point. The abatement cost at this emission charge increases to $150 million for the most stringent technology, corresponding to a 38% decrease in emissions.

The abatement cost is the cost to firms of abatement net of the emission charge paid by them to the regulatory authority, which is returned to firms as a lump-sum transfer. The abatement cost represents the cost of inputs required for abatement. When all abatement possibilities have been exhausted, the abatement cost curve flattens out. In this range, the amount of abatement is virtually constant and so is the cost of abatement. The maximum abatement cost ranges from about $160 million for the least stringent technology, to $230 million for the most stringent technology.

Figure 3.2c shows how the NO_x emission charge affects regional value added for a much wider range of emission charges, up to $350,000 per ton, far above the marginal cost of control for most abatement technologies. In this figure, given the wider range, the three lines for different technologies are virtually congruent. The highest marginal cost is about $38,000 per ton for all but one source category and technology, which has a marginal cost of $407,000 per ton. The impact on regional value added for emissions charges within the range of abatement costs is small. Regional value added

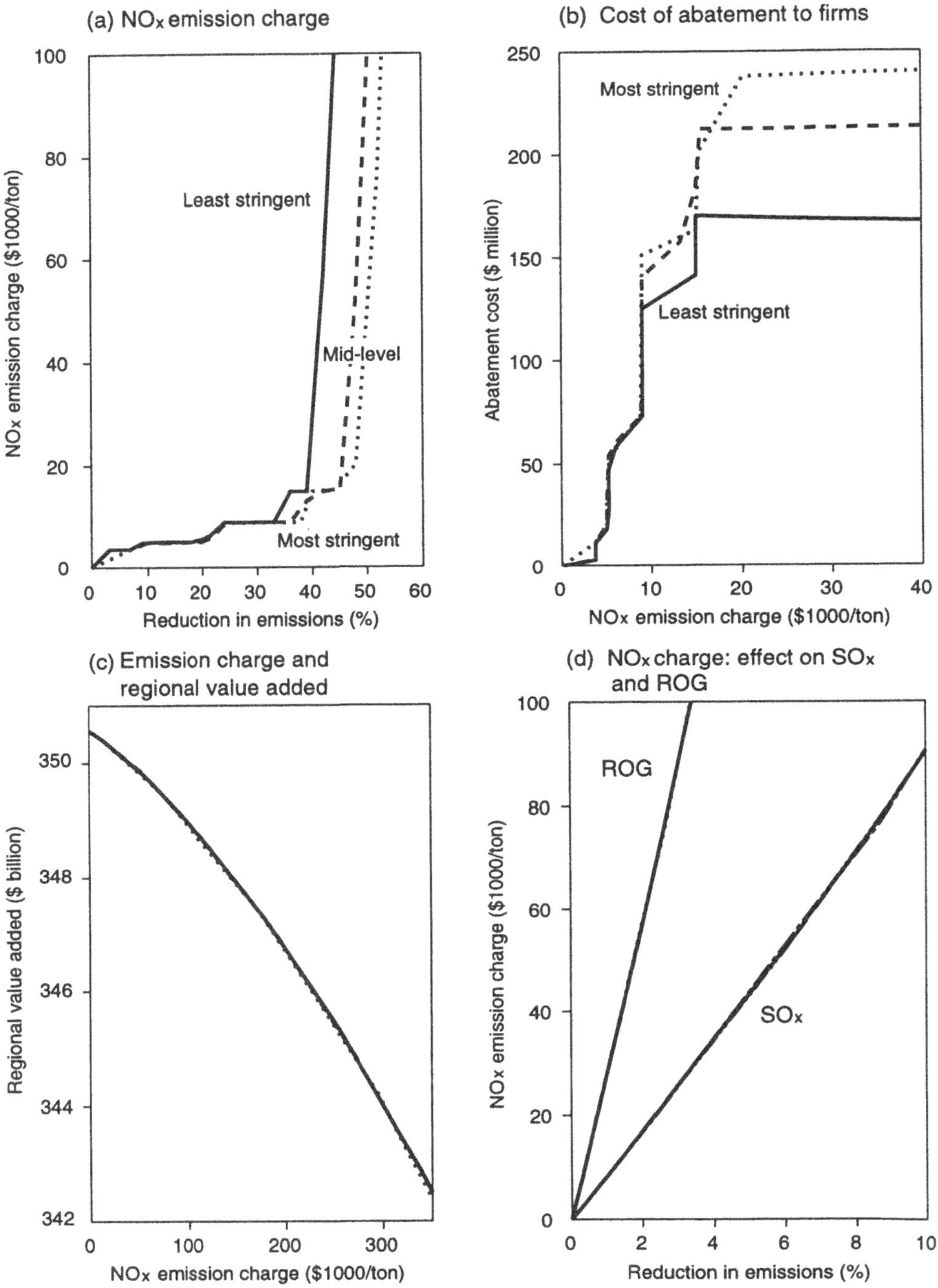

Figure 3.2. NO_x emission charge.

falls from \$350.57 billion at a zero emission charge, to \$350.02 billion at an emission charge of \$40,000 per ton, well above the marginal cost of all but one technology. Thus, with a charge of around \$20,000 per ton, one achieves about 40% reduction in emissions of NO_x with little impact on regional value added. After that, the no-abatement model starts to kick in, with a much larger impact on regional value added for incremental reductions in NO_x.

Figure 3.2d shows how the NO_x emission charge affects emissions of other pollutants. For emission charges up to \$40,000 per ton, the impact on ROG and SO_x is small. This result is to be expected because the impact on production structure of such an emission charge is small. At this emission charge, the corresponding *ad valorem* taxes are small, reaching 6%–7% only for the utilities, air transportation, and glass sectors. Finally, notice that in both *Figures 2c* and *2d*, the curves for the least stringent, mid-level and most stringent technologies coincide extremely closely so that the adoption of more stringent abatement technologies will not affect the conclusions drawn above.

(b) ROG Emission Charge

In the case of ROG, the emission charge required for a similar reduction in emissions is substantially higher than for NO_x because the abatement technologies have higher costs and because abatement possibilities are extremely limited in some major ROG-producing sectors (*Table 3.2*). The transition from the region of low emission charges to that of high emission charges (*Figure 3.3a*) is more gradual than in the case of NO_x. One reason for this result is that the range of abatement costs is broader so that there are several high-cost technologies (with marginal costs around \$100,000 per ton) which are adopted well after the lower-cost technologies have been put to use. In addition, there are many more control categories for ROG than for NO_x, which, combined with the greater dispersion in control costs, gives an emission cost curve which is less flat. The maximum cost of abatement to firms ranges from about \$250 million to \$370 million, depending on the technology in use. Notice that the abatement cost curves begin to decline when the emission charge goes above \$100,000 (*Figure 3.3b*). At these high levels of emission charge, the amount of ROG being produced before abatement begins to fall quite noticeably because the charge is causing a reduction in output in the polluting sectors. Since there is less output and fewer emissions to be abated, the cost of abatement falls.

Figure 3.3c shows the impact of the ROG emission charge on regional value added. At an emission charge of \$120,000 per ton, regional value

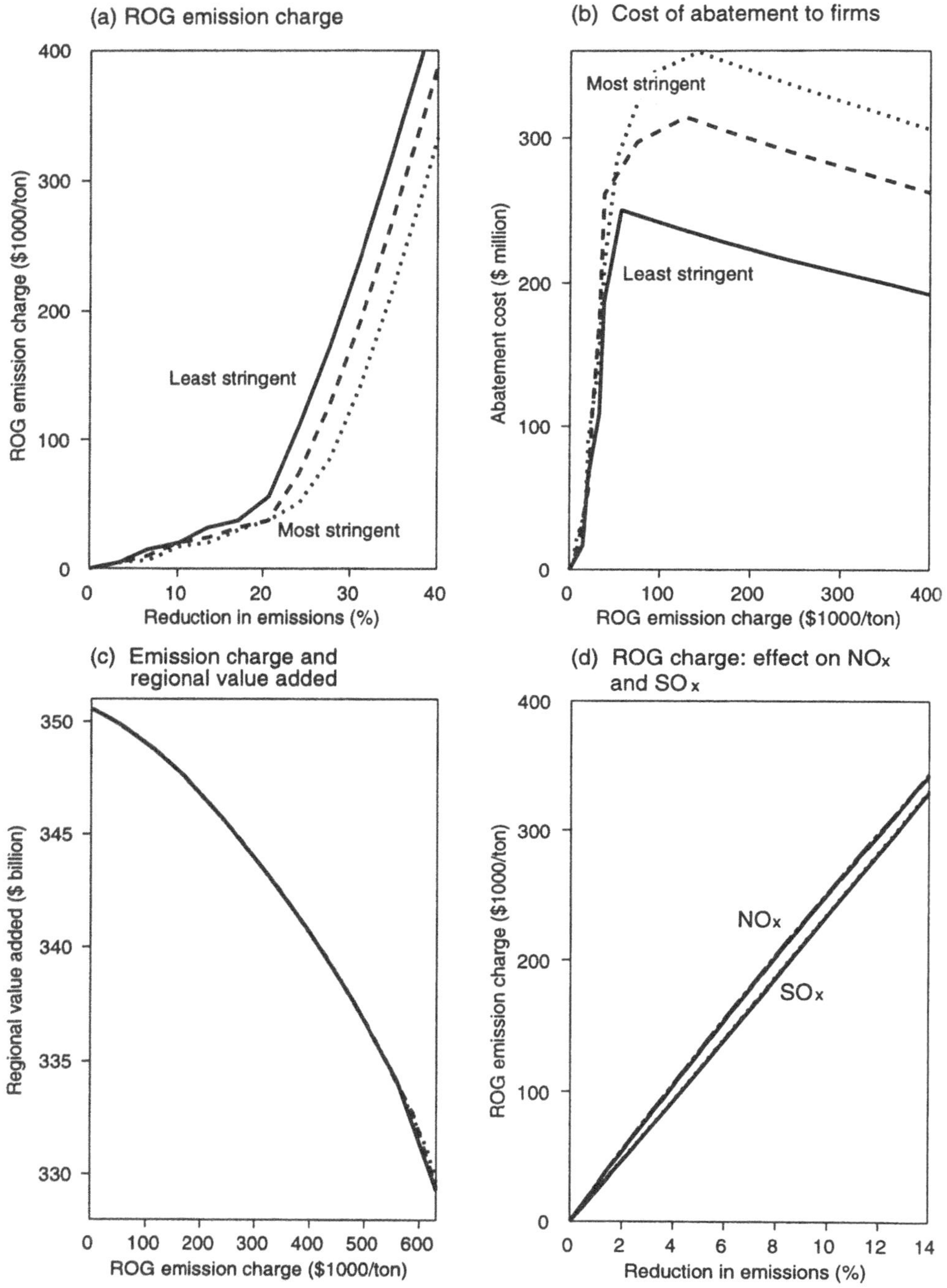

Figure 3.3. ROG emission charge.

added falls from \$350.57 billion to \$348.60 billion at a zero emission charge. Of course, this decrease is larger than in the case of the NO_x emission charge of \$40,000 per ton considered above because the corresponding *ad valorem* tax rates are higher – 17% for wood products and air transportation, and 12% for oil and gas extraction. As in the case of the NO_x emission charge, the effect of the ROG emission charge on NO_x and SO_x emissions is small so long as the ROG emission charge is not extremely high. Again, the lines for the different abatement technologies are virtually the same, given that the scale is well beyond the range where abatement is fully utilized.

(c) SO_x Emission Charge

In the case of sulfur oxides, the abatement technologies all have low marginal costs of between \$1,000 and \$15,000 per ton, except for two with costs around \$60,000 per ton.[4] As can be seen from *Figures 4a* and *4b*, reductions of 30%–35% are achieved at low cost. For reductions approaching 40%, the emission charge increases steeply. Abatement costs range from around \$20 million for the least stringent technology to \$48 million for the most stringent technology. Since the abatement costs are low and the SO_x emission coefficients (in tons per million dollars of output) are small compared to those for the other pollutants, the effect on regional value added is small (*Figure 3.4c*). At an emission charge of \$80,000 per ton, the *ad valorem* tax rates are at most 3% (for petroleum, utilities, and glass and cement sectors) and regional value added falls only by \$ 0.24 billion. Again, the impact on ROG and NO_x is small.

3.5 Conclusions

The results show that regional CGE models are useful for modeling air pollution control in some detail. We have succeeded in integrating a nonlinear programming model of abatement with a multisectoral, general equilibrium, simulation model of the LA regional economy. The simulations described here show how different levels of abatement can be attained using emission charges and at what cost to firms and to the regional economy. The key empirical finding is that for levels of abatement that are technically feasible – i.e., for which technologies exist – the costs of abatement to firms are relatively small (of the order of a few hundred million dollars, which is less than 1% of sales) and the reduction in regional value added is also small (of the order of 1%–2%). After abatement possibilities are exhausted, it is necessary to institute much higher emission charges in order to change the

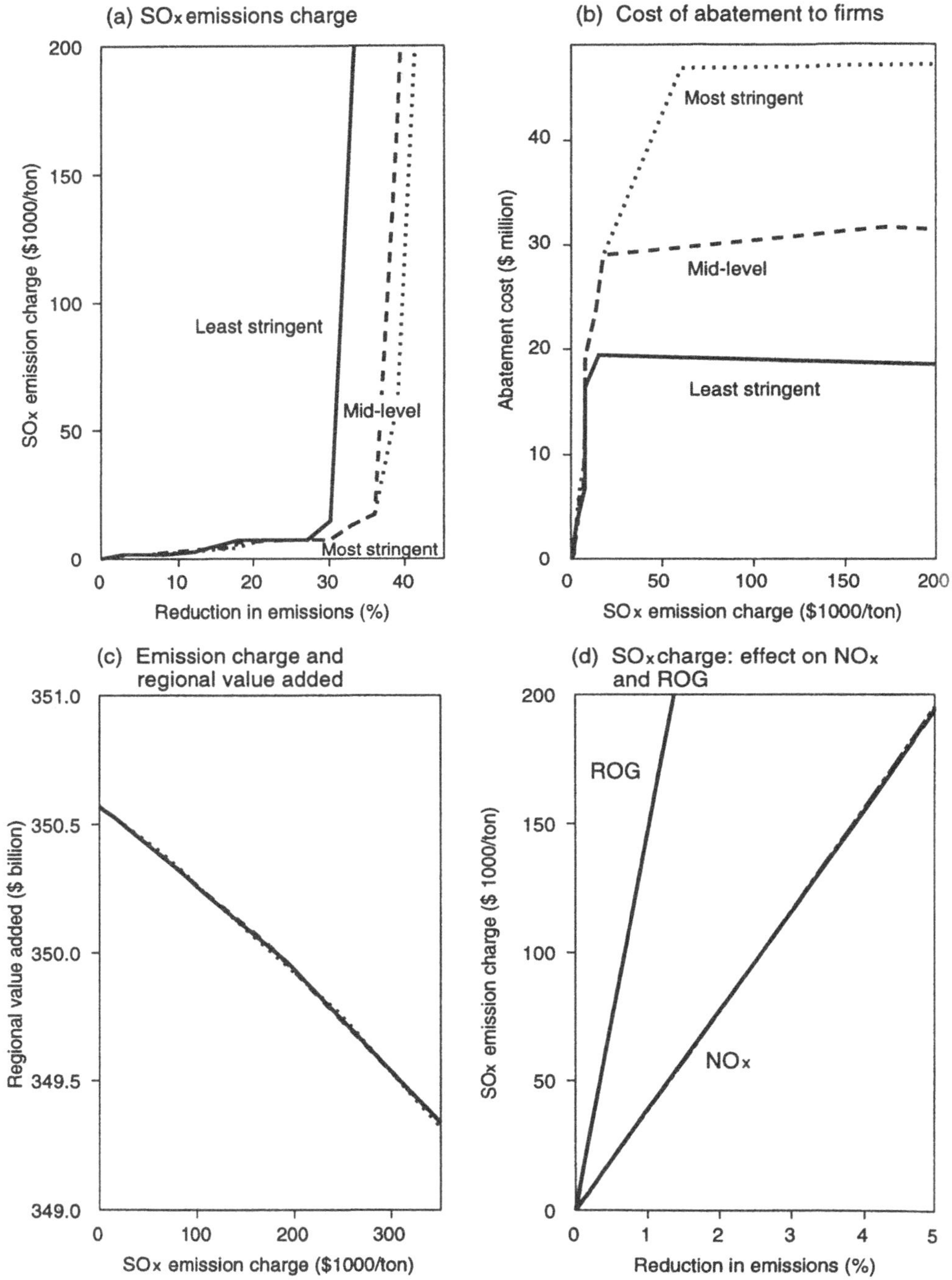

Figure 3.4. SO_x emission charge.

structure of production in the region away from polluting sectors. The costs of this approach in terms of lower regional product are much greater. While the model represents the complexity of regulatory policies in a stylized way, these key findings are likely to be robust.

The model results suggest a number of areas for further refinement. An important step would be to improve the treatment of SO_x abatement. It is apparently the case that for several source categories, SO_x emissions can be reduced considerably by changing from sulfur-laden fuels to natural gas. The cost of this changeover at current prices is small. Modeling this substitution possibility would require changing sectoral intermediate demands (the input–output coefficients) to reflect the substitution of natural gas for other fuels and reducing the SO_x emissions in the base year for the affected source categories. There are no conceptual or serious modeling difficulties involved here, but this step would require substantial amounts of data.

The modeling of SO_x abatement in the utilities sector also needs to be changed to take into account the fact that plant-specific data on abatement technology are available. Since this approach would require working at the level of individual generating plants, data on base-year emissions for each plant would be needed. It might also be necessary at this level of detail to take into consideration the regulated nature of this sector.

The economic data used in the model are for 1990, while the emission data are for 1989. It would be a good idea to work with data for a single, recent year, if possible. It would also be worthwhile to disaggregate the economic data. For example, one might disaggregate labor by skill type to capture distributional effects on the labor markets. Adding more sectoral detail would permit closer integration of engineering and economic data.

Finally, the model abstracts from many of the complexities of the proposed regulations for the RECLAIM and New Source Review (NSR) programs. For example, the programs exclude small producers, who are significant in aggregate economic activity in the region. There is always a trade-off in empirical modeling between simplification and realism. Our stylized model can certainly be expanded to capture more of the institutional details of the programs, but the question is whether the additional detail matters to the essential results. The model captures the major links between polluting and nonpolluting sectors and the important features of abatement. Adding more sectors or incorporating more institutional detail, while refining the analysis, also adds to the model's complexity. The modeler's art is to achieve sufficient realism for policy relevance while retaining transparency, so that the mechanisms at work are clearly visible and understandable.

Acknowledgments

Funding for this project was provided by the California Energy Commission. Any views expressed in this chapter are those of the authors and do not necessarily represent the views or policies of the funding institution. We would like to thank participants at a number of seminars, the editors, and two anonymous referees for very helpful comments.

Notes

[1] Aggregate Social Accounting Matrices (SAMs) for the counties are available only for 1982. The 1982 Los Angeles basin SAM from IMPLAN was updated to 1990 by using the IMPLAN input–output data for 1982 and 1990 and growth rates for various state-level aggregates (from the California Statistical Abstract).

[2] The model is formulated in such a way as to allow great flexibility in changing the sector definitions. A 39-sector version also exists, but only the results from the 16-sector version are presented here.

[3] The CEC's own data on sectoral output for the Los Angeles region are likely to be more reliable than those from the IMPLAN database. It would be worthwhile to re-base the economic data to 1989 using the CEC's sectoral output data for 1989. However, these data would then have to be reconciled to the IMPLAN data for other variables.

[4] The results for SO_x abatement must be treated as preliminary in nature because input substitution (natural gas for sulfur-containing fuels), which is not modeled here, offers a low-cost and potentially effective means of abatement. In addition, for the utility sector, the CEC data set provides cost data separately for three different electricity generating plants. These cost data could not be used since information on the base year SO_x emissions from these plants was unavailable. Instead, an average of these costs has been used.

References

Baumol, W., and Oates, W.E., 1971, The Use of Standards and Prices for Protection of the Environment, *Swedish Journal of Economics* **73**(1):42–54.

Bergman, L., 1988, Energy Policy Modeling: A Survey of General Equilibrium Approaches, *Journal of Policy Modeling* **10**(3):377–399.

Bergman, L., 1990, Energy and Environmental Constraints on Growth: A CGE Modeling Approach, *Journal of Policy Modeling* **12**(4):671–691.

Bergman, L., 1991, General Equilibrium Effects of Environmental Policy: A CGE Modeling Approach, *Environmental and Resource Economics* **1**(1):43–61.

Brooke, A., Kendrick, D., and Meeraus, A., 1988, *GAMS, A User's Guide*, Scientific Press, Redwood City, CA, USA.

Dervis, K., de Melo, J., and Robinson, S., 1982, *General Equilibrium Models for Development Policy*, Cambridge University Press, New York, NY, USA.

Hahn, R.W., and Hester, G.L., 1989, Marketable Permits: Lessons for Theory and Practice, *Ecology Law Quarterly* **16**(2):361–406.

Hazilla, M., and Kopp, R.J., 1990, Social Cost of Environmental Quality Regulations: A General Equilibrium Analysis, *Journal of Political Economy* **98**(4):853–873.

Hollenbeck, K., 1979, The Employment and Earnings Impacts of the Regulation of Stationary Source Air Pollution, *Journal of Environmental Economics and Management* **6**:208–221.

Jorgenson, D.W., and Wilcoxen, P.J., 1990, Environmental Regulation and US Economic Growth, *RAND Journal of Economics* **21**(2):314–340.

Jorgenson, D.W., and Wilcoxen, P.J., 1990, Intertemporal General Equilibrium Modeling of US Environmental Regulation, *Journal of Policy Modeling* **12**(4):715–744.

Leontief, W., 1970, Environmental Repercussions and the Economic Structure: An Input–Output Approach, *Review of Economics and Statistics* **52**(2):262–271.

Malik, A.S., 1990, Markets for Pollution Control when Firms are Noncompliant, *Journal of Environmental Economics and Management* **18**:97–106.

Milliman, S.R., and Prince, R., 1989, Firm Incentives to Promote Technological Change in Pollution Control, *Journal of Environmental Economics and Management* **17**:247–265.

Montgomery, D.W., 1972, Markets in Licenses and Efficient Pollution Control Programs, *Journal of Economic Theory* **5**:395–418.

Robinson, S., 1989, Multisectoral Models, in H.B. Chenery and T.N. Srinivasan, eds., *Handbook of Development Economics*, North Holland, Amsterdam, Netherlands.

Robinson, S., 1990, *Pollution, Market Failure, and Optimal Policy in an Economy-wide Framework*, Working Paper No. 559, Department of Agricultural and Resource Economics, University of California, Berkeley, CA, USA.

South Coast Air Quality Management District, October 1993, RECLAIM: The Regional Clean Air Incentives Market: Final Volume 1, Diamond Bar, CA, USA.

Tietenberg, T.H., 1985, *Emissions Trading*, Resources for the Future, Inc., Washington, DC, USA.

Appendix: The Los Angeles Basin CGE Model

Prices block

$$P_i^m = P_i^{wm} ER(1 + t_i^m) \tag{1}$$
$$P_i X_i = P_i^d X_i^{xd} + P_i^m M_i \tag{2}$$
$$P_i^x X_i^d = P_i^d X_i^{xd} + P_i^e E_i \tag{3}$$
$$P_i^e = P_i^{we} ER(1 + t_i^e) \tag{4}$$
$$P_i^{va} = P_i^x (1 - t_i) - \sum_j A_{ji} P_j - A_i^{nc} ER \cdot P_i^{nc} - t_i^{pol} \tag{5}$$
$$\overline{P} = \sum_i w_i P_i^x \tag{6}$$
$$P_i^k = \sum_j B_{ji} P_j \tag{7}$$

Production block

$$X_i^d = A_i^d \prod_{i,f} FD_{if}^{\nu_{if}} \tag{8}$$
$$W_f \cdot WD_{if} \cdot FD_{if} = \nu_{if} \cdot P_i^{va} \cdot X_i^d \tag{9}$$
$$ID_i = \sum_j A_{ij} X_j^d \tag{10}$$
$$X_i^d = CET(X_i^{xd}, E_i), \quad i \in I_E \tag{11}$$
$$X_i^d = X_i^{xd}, \quad i \in \overline{I}_E \tag{12}$$
$$E_i = X_i^{xd} \cdot f^e(P_i^e / P_i^d), \quad i \in I_E \tag{13}$$
$$X_i = CES(X_i^{xd}, M_i), \quad i \in I_M \tag{14}$$
$$X_i = X_i^{xd}, \quad i \in \overline{I}_M \tag{15}$$
$$M_i = X_i^{xd} \cdot f^m(P_i^d / P_i^m), \quad i \in I_M \tag{16}$$
$$YI_{ent} = YF_K + GT_{ent} + T_{pol} - C_{ctl} - S_{ent} - T_{ent} - S_{dep} \tag{17}$$

Incomes block

$$YF_f = \sum_i W_f \cdot WD_{if} \cdot FD_{if} + ER \cdot YF_f^w \tag{18}$$
$$YI_{lab} = YF_L - T_{soc} \tag{19}$$
$$T_{ind} = \sum_i t_i P_i^x X_i^d \tag{20}$$
$$T_{soc} = t_{soc} YF_L \tag{21}$$
$$T_{ent} = t_{ent} \left(YF_K - S_{dep} + GT_{ent} \right) \tag{22}$$
$$S_{dep} = \sum_i d_i P_i^k FD_{iK} \tag{23}$$
$$S_{ent} = s_{ent} \left(YF_K + GT_{ent} - T_{ent} - S_{dep} \right) \tag{24}$$
$$S_h = \sum_h s_h YH_h (1 - t_h) \tag{25}$$
$$GR = T_m + T_{ind} + T_h + T_{soc} + T_{ent} + ERK_w \tag{26}$$
$$S = S_h + S_g + S_{dep} + S_{ent} + ERS_w \tag{27}$$

Expenditures block

$$P_i CD_i = \sum_h w_{ih}^c (1 - s_h)(1 - t_h)(1 - w_h^{nc}) YH_h \tag{28}$$

$$\sum_i P_i GD_i = \left(1 - w_g^{nc}\right) GD_{tot} \tag{29}$$

$$GR = GD_{tot} + S_g + GT_{ent} + GT_h \tag{30}$$

$$\Delta S_i = \delta_i X_i^d \tag{31}$$

$$I_f = (1 - w_I^{nc})(I - \sum_i P_i \Delta S_i) \tag{32}$$

$$P_i IV_i = w_i^I I_f \tag{33}$$

$$\begin{aligned} M_{nc} = {} & w_g^{nc} GD_{tot} + w_I^{nc} (I - \textstyle\sum_i P_i \Delta S_i) \\ & + \textstyle\sum_h w_h^{nc} (1 - s_h)(1 - t_h) YH_h + ER \sum_i P_i^{nc} A_i^{nc} X_i^d \end{aligned} \tag{34}$$

Market clearing

$$X_i = ID_i + CD_i + GD_i + IV_i + \Delta S_i + AD_i \tag{35}$$

$$\sum_i FD_{if} = FS_f \tag{36}$$

$$\sum_i P_i^{wm} M_i + \frac{M_{nc}}{ER} = \sum_i P_i^{we} E_i + \sum_f YF_f^w + S_w + R + K_w \tag{37}$$

$$GDP = \sum_i P_i^{va} X_i^d + T_{ind} + T_m + T_{pol} \tag{38}$$

Abatement block

$$Q_{ip}^0 = \gamma_{ip} X_i^d \tag{39}$$

$$AB_{ikpt} = y_{kpt} Q_{ip}^0 \theta_{ikp} \alpha_{kpt} \tag{40}$$

$$Q_p = \sum_i Q_{ip}^0 - \sum_{i,k,t} AB_{ikpt} \tag{41}$$

$$Q_p \leq \overline{Q}_p \tag{42}$$

$$t_i^{pol} = \sum_p \tau_p \gamma_{ip} \left(1 - \sum_{k,t} y_{kpt} \theta_{ikp} \alpha_{kpt}\right) \tag{43}$$

$$T_{pol} = \sum_i t_i^{pol} X_i^d \tag{44}$$

$$P_{kpt}^{ab} \cdot y_{kpt} \leq \tau_p \cdot y_{kpt} \tag{45}$$

$$P_{kpt}^{ab} (1 - y_{kpt}) \geq \tau_p (1 - y_{kpt}) \tag{46}$$

$$\Omega = \sum_{i,k,p,t} AB_{ikpt} \tag{47}$$

$$C_{ctl} = \sum_{i,k,p,t} P_{kpt}^{ab} AB_{ikpt} \tag{48}$$

$$P_{kpt}^{ab} = \sum_i b_{ikpt} P_i \tag{49}$$

$$AD_i = \sum_{k,p,t} b_{ikpt} \sum_j AB_{jkpt} \tag{50}$$

Parameters

A_i^d Constant in production function
A_i^{nc} Noncompetitive import coefficient in sector i
A_{ij} Input–output coefficient
B_{ij} Capital composition matrix
ν_{if} Production function parameter
α_{kpt} Emission reduction factor for pollutant p from source k using technology t
b_{ikpt} Input requirements from sector i for abatement technology t for pollutant p and source category k
d_i Depreciation rate in sector i
δ_i Stock change coefficient in sector i
γ_{ip} Emission coefficient, tons of pollutant p per unit output in sector i
s_{ent} Enterprise savings rate
s_h Household savings rate
t_{ent} Enterprise tax rate
t_h Household tax rate
t_i Indirect tax rate in sector i
t_{ocs} Social security tax rate
θ_{ikp} Share of source category p in emissions from sector i for pollutant p
w_I^{nc} Share of investment demand met by noncompetitive imports
w_g^{nc} Share of government demand met by noncompetitive imports
w_h^{nc} Share of consumption demand met by noncompetitive imports
w_i Weights for producer price index
w_i^I Shares by sector of origin of investment demand
w_{ih}^c Budget shares for consumer demand

Sets and other miscellaneous notations

f Factors, capital (K) and labor (L)
i Sectors
p Pollutants
k Emission source categories
t Abatement technologies, $t = 1, 2, 3$
I_E Set of sectors with positive exports
I_M Set of sectors with positive imports
$\overline{I}_E$ Set of sectors with no exports
$\overline{I}_M$ Set of sectors with no imports
f^e First-order condition for CET function for exports
f^m First-order condition for CES function for imports

Exogenous variables

ER	Exchange rate
FS_f	Factor supply for factor f
GD_i	Government demand for sector i
GT_{ent}	Government-to-enterprise transfer
GT_h	Government-to-household transfer
I	Total investment
K_w	Foreign borrowing by government
P_i^e	Price of exports in sector i
P_i^m	Price of imports in sector i
P_i^{nc}	Price of noncompetitive imports in sector i
P_i^{we}	World price of exports in sector i
P_i^{wm}	World price of imports in sector i
$\overline{Q}_p$	Target for total emissions of pollutant p
R	Remittances to households from rest of the world
WD_{if}	Factor price differential in sector i for factor f
YF_f^w	Factor income for factor f from abroad

Endogenous variables

AB_{ikpt}	Abatement in sector i for pollutant p from source k using technology t
AD_i	Demand for inputs from sector i arising from abatement activities
CD_i	Consumer demand for sector i
ΔS_i	Change in stocks for sector i
C_{ctl}	Abatement costs
E_i	Exports in sector i
FD_{if}	Factor demand for factor f in sector i
GDP	Gross domestic product
GD_{tot}	Total government demand
ID_i	Intermediate demand for sector i
IV_i	Investment demand for sector i
I_f	Fixed investment
M_i	Competitive imports in sector i
M_{nc}	Noncompetitive imports
Ω	Objective function
$\overline{P}$	Aggregate price level
P_i	Price of composite commodity in sector i
P_i^d	Price of domestic sales in sector i
P_i^k	Price of capital in sector i

P_i^{va}	Price of value added in sector i
P_i^x	Average price of output in sector i
P_{kpt}^{ab}	Cost of abatement for pollutant p from source k using technology t
Q_{ip}^0	Pre-abatement emissions of pollutant p in sector i
Q_p	Total emissions (post-abatement) of pollutant p
S	Total savings
S_{dep}	Depreciation
S_{ent}	Enterprise savings
S_g	Government savings
S_h	Household savings
S_w	Foreign savings (current account deficit)
T_{ent}	Taxes on enterprises
T_h	Taxes on households
T_{ind}	Indirect taxes
T_{pol}	Total emission charge revenue
T_{soc}	Social security taxes
W_f	Average price for factor f
X_i	Composite commodity in sector i
X_i^{xd}	Domestic sales in sector i
X_i^d	Domestic production in sector i
YF_f	Factor income for factor f
YH_h	Household income for household group h
YI_{ent}	Enterprise income
YI_{lab}	Labor income
t_i^{pol}	Pollution tax in sector i
τ_p	Emission charge per ton of pollutant p
y_{kpt}	Binary variable for abatement technology t, source category k, and pollutant p

Part II

National Experiences

Chapter 4

Developments in the Use of Economic Instruments in OECD Countries

Hans Opschoor
Department of Spatial Economics
Free University, Amsterdam, The Netherlands

Abstract

For the period 1987–1993, developments in the environmental policies of OECD countries with respect to the use of economic instruments are compared and the differences analyzed. The focus is on applications in the field of air pollution policies. The comparison is made on the basis of two surveys. To complete the descriptive part, a brief survey is also presented of currently discussed and recently introduced economic instruments. A description of economic instruments as such and a review of rationales for employing economic and financial incentives precede this analysis. The analysis shows that the use of economic instruments has indeed increased since 1987, but the development has not been spectacular. Possible explanations for this are presented. Also, some types of instrument have advanced more than others and the changes differ from one set of countries to another. Product charges (including air pollution-related ones) have become more widely used, especially in Scandinavian countries. Moreover, growing attention is being paid to the use of economic instruments at the international level. The incentive impacts of economic (and other) instruments appear to have received relatively little empirical attention, even though these are an important policy-relevant feature in instrument choice.

Key words: economic instruments, environmental policy, environmental economics, air pollution, air pollution policies.

4.1 Introduction and Background

Economic instruments have been hailed by economists since the early days of environmental economics, but they have had a limited impact on environmental policy. Since the Brundtland Report (WCED, 1987) a general interest has been aroused in what the report called "merging environment and economics in decision making", and economic instruments were regarded as one potential way of achieving that. One may wonder to what degree this has altered the mixes of instruments used in environmental policies. This chapter describes and analyzes developments in the use of economic instruments, and attempts to look ahead to discern what possible future they may have. The focus is on instruments used in air pollution policies, but this is placed in the wider context of environmental policy.

Narrowly defined, policy instruments could be labeled "economic" if they have intended and/or effective incentive impacts. Economic instruments, thus defined, have indeed been applied in environmental policies of industrialized counties for some time, but they have been rather rare. More broadly defined, economic incentives include a number of revenue raising or financial instruments, and the latter definition is the more common one. In this chapter we deal with economic instruments in the broader sense. In Section 4.2 the concept of economic instruments for environmental policies is discussed and rationales for deploying them are reviewed. Section 4.3 provides an empirical description and a qualitative analysis of the use of economic instruments. A comprehensive overview is presented of the present uses (in OECD member countries) of the following types of economic instruments of environmental policy: charges on emissions; charges on products; deposit–refund systems; tradable permits; and enforcement incentives. The issue of the incentive impact of these instruments is introduced, but it appears that assessments (in an international comparative context) of such impacts require data that are not yet available. Section 4.3 also presents some developments over time in the use of economic instruments. It appears that some types of economic instruments have been put into practice much more often than others, and that there are large differences between countries in the way they are used. These differences may not be the results of coincidence, but probably reflect institutional differences between countries and regions, that must be taken into account and which may have significance in considerations on international harmonization. A discussion leading to an outlook on possible future uses, and some conclusions, is given in Section 3.4.

This chapter is based on two surveys conducted for the OECD on the uses of economic instruments in its member countries (Opschoor and Vos, 1989; Vos *et al.*, 1994). The results of these two studies allow for some longitudinal analysis, although, of course, the significance of a one five-year period is likely to be limited.

4.2 Economic Instruments in Environmental Policy in OECD Countries

4.2.1 Introduction

Economic instruments are considered to be capable of providing incentives for technological innovation and behavioral change, and to offer good prospects for achieving environmental objectives in a cost-effective manner (Baumol and Oates, 1975). According to an official OECD statement (OECD, 1991, par. 20) they improve the allocation and efficient use of natural and environmental resources. At UNCED a number of economically significant positions were accepted, including the need to internalize environmental costs, the precautionary approach to environmental change, and the use of economic instruments (in the Rio Declaration and *Agenda 21*).

In Section 4.2.2 economic instruments are first positioned in relation to (environmental) policy instruments in general. Section 4.2.3 presents a practical classification (as applied in the OECD surveys, for example) for distinguishing between five instrument categories: charges on emissions, charges on products (including tax rebates), deposit–refund systems, market creation (including tradable permits and quotas), and enforcement incentives. Finally, in Section 4.2.4 we review the debate on the theoretical and practical significance of economic instruments for environmental policy. Readers who are reasonably familiar with economic instruments in general are advised to proceed to Section 4.2.4 immediately.

4.2.2 Environmental policy instruments

In seeking to ensure the sustainable use of environmental resources and the maintenance of stricter levels of environmental quality, environmental policy can make use of (mixes of) two basic strategies: (a) to engage in public projects and programs aimed at preventing, compensating for, and eliminating environmental degradation, or at providing substitutes for traditional behavioral patterns; and (b) influence the decision making process at the micro-level, i.e., that of the environmentally relevant (economic) agents such as consumers, producers, and investors. The second strategy is discussed in some detail below.

Rational decision makers will base their decisions about their activities on a comparison of the benefits and costs to them of the available and perceived options. These decisions can be influenced in three basic ways: (1) alteration of the set of options open to agents; and (2) alteration of the cost and/or benefits relevant to agents; and (3) alteration of the priorities and significance agents attach to environmental change (i.e., altering the structure of agents' preferences or objective functions; Opschoor, 1990). Route (1) involves providing new alternatives or forbidding (or licensing) old ones. Typically, this has been the route followed by environmental policy in most industrialized countries, the so-called "command-and-control" approach. Instruments used in this approach have included a wide range of "direct" regulations (i.e., regulations that directly influence behavior from an external "leverage point"), such as standards, bans, permits, zoning, quota, use restrictions, etc. These instruments present cases of direct regulation. Route (2) leads to the application of economic incentives, via markets or otherwise. The motivation relied upon here is that if environmentally more appropriate behavior is made more rewarding in the eyes of the agent involved, then attitudes and behavior will "automatically" shift in favor of these socially more desirable alternatives. Options can be made more or less (financially or economically) attractive by applying charges or levies, granting subsidies, implementing tax differentiation, etc. (see below for a more complete list). Such instruments are referred to below as economic instruments. In this way environmental concerns can in a certain sense be "internalized" by altering the agent's context rather than the agent's value structure or preferences. Route (3) entails approaches such as: education, information extension, and training, but also social pressure, negotiation and other forms of "moral suasion". Here the mechanisms include: (i) improving the information base of the agent, (ii) changing perceptions and priorities within the agent's preference structure, and (iii) expressing the priorities and preferences of other relevant parties in a negotiating context.

The notion of an "economic" instrument has come to mean different things in different contexts. Instruments can be labeled "economic" in a strict sense, when they are intended to affect estimates of the costs and benefits of alternative actions open to economic agents. The effect would be that decision making and behavior are influenced in such a way that from the available options the ones that are chosen are those that lead to a socially more desirable situation than in the absence of the instrument; agents are assumed to adapt to the introduction of the instrument by preferring the least-cost option. If economic agents indeed behave in such a way, then environmental agencies could determine their environmental objectives and, through such incentive-based instruments, could manipulate the economic agents to such a degree that these environmental objectives would be attained in a cost-effective way. For example, emissions trading schemes may lead to least-cost approaches to attaining certain ambient quality standards or overall discharge levels, if those under the scheme are interested in cost minimization or profit maximization. Such instruments are included in the above definition of economic instruments. Direct regulation often also has a financial or monetary component attached to it. For instance, when a permit is issued, firms will have to bear the costs of complying with it. In some cases a regulation is accompanied by charges that have no intended impact on behavior, but in reality do affect it. An example is the Dutch water pollution charge, which was intended to recover the costs of collective treatment but which achieved such high levels that Dutch industry to a large degree preferred to resort to private, on-plant treatment so that they would not have to pay the charge. Yet they do not fall within the strict definition of economic instruments. Many actual charge systems would thus not qualify as economic instruments under this definition. However, many wider definitions include instruments with direct financial components. In the OECD reviews of the use of economic instruments a pragmatic and fairly broad approach has been taken (see Section 4.2.3).

4.2.3 Economic instruments: A classification

Four elements had to be present for an instrument to be selected for this survey. First, alternative behavioral options had to be available to an economic agent. Second, there would have to be involvement of government (related) authorities. Third, this authority would have to have an intention of (directly or indirectly) maintaining or improving environmental quality by applying the instrument. Finally, an impact on levels of costs or benefits of alternative behavioral options would have to emanate from the application of the instrument. The existence of financial stimuli (either negative or positive) is not one of the common elements, although it is almost always present. Apart from subsidies, five main categories of economic instruments can be distinguished: charges on emissions, charges on products, deposit–refund systems, market creation, and financial enforcement incentives.

Charges

To some extent, charges may be regarded as a "price" to be paid for pollution (in a welfare economic setting, where the social costs of pollution are considered), although in fact they very often have other functions, such as providing a disincentive to environmentally damaging behavior, or reflecting the costs of pollution or waste treatment services. Polluters have to pay for their implicit claim on environmental "services", which thereby enters at least in some part into private cost–benefit calculations. Charges may have an incentive impact and a revenue- (or fund-) raising impact. The incentive impact of charges depends on the cost and price changes brought about by the charge. In many cases, charges mainly have a revenue-raising impact: they are too low to have an incentive impact and the revenues are intended for collective treatment, for research on new abatement technologies, or for subsidizing new investment. The following types of charges are reviewed:

- *Charges on emissions or effluent charges* are to be paid on discharges into the environment and are based in principle on (a proxy for) the quantity and/or quality of discharged pollutants. Emissions charges may take the form of "user charges": these are payments for the costs of collective or public treatment of effluent. Tariffs may be uniform or they may differ according to the amount of effluent treated. Such user charges are very much like the price of any public good. Otherwise, i.e., outside the context of the user of treatment facilities, emission-based charges may reflect incentive aims or other revenue-raising objectives that are not directly related to the financing of the specific pollutant-treatment facilities upon which the charge is based.
- *Product charges* are charges on products that pollute in the manufacturing or consumption phase, or for which a disposal system has been organized. Product charges can be based on some characteristic of a product (e.g., on the sulfur content of mineral oil) or on the product itself (e.g., a mineral oil charge). One form that product charges may take in economic practice is that of tax differentiation leading to more favorable prices for "environmentally friendly" products, and vice versa. The sole purpose of tax differentiation is its incentive impact and this instrument often operates budget-neutrally, while product charges could also have a revenue-raising goal. Administrative charges such as control and authorization fees and payments for administrative services, such as the registration of certain chemicals, or the implementation and enforcement of regulations, are not analyzed here.

Deposit–Refund Systems

In deposit–refund systems a surcharge is added to the price of potentially polluting products. When pollution is avoided by returning these products or their residuals to a collection system, the surcharge is refunded. The best known deposit–refund system is that applied to beverage packaging.

Market Creation

One can speak of market creation when economic agents are given "rights" for actual or potential pollution and where they can trade their "pollution rights" or their process residuals (recycled materials). Several types exist:

- *Emissions trading.* In this approach, dischargers operate under some multi-source emission limit and trade is allowed in permits adding up to that limit. Such systems can also operate in cases of single-source permits: if a discharger releases less pollution than its limit allows, the firm can sell or trade the differences between its actual and allowable discharges to another firm, which then has the right to release more than its initial limit allows. Trades can take place within a plant, within a firm or among different firms; in the past, most trades have occurred within a plant or a firm.
- *Trading in resource extraction quotas.* Resource extraction concessions can be issued, and trading in these could be allowed. Examples include timber concessions in the forestry sector, or in fish catch quotas. Such quota trading schemes are not considered here.

There may be other types of market creation that in principle could be regarded as economic instruments of environmental policy. These include price interventions (subsidies in case market prices fall below certain levels) or *ex ante* price guarantees that create or facilitate the continued existence of a market, and liability of polluters for environmental damage or clean-up costs (where the liability gives rise to new insurance activities). These issues are also beyond the scope of this chapter.

Enforcement Incentives

(Financial) enforcement incentives are sometimes regarded as legal rather than economic instruments: noncompliance is "punished" either *ex ante* (by requesting a payment that is returnable upon compliance) or *ex post* (by imposing a fine when noncompliance occurs). However, enforcement incentives may provide an economic rationale for compliance, when noncompliance is a seriously considered decision alternative. Two types of enforcement incentives are therefore included in this overview:

- Noncompliance fees imposed when polluters do not comply with certain regulations and where the rates are more or less proportional to noncompliance benefits reaped or environmental damage inflicted.
- Performance bonds, or payments to authorities in expectation of compliance with imposed regulations. Refunds are given when compliance has been achieved.

In Section 4.3 we attempt to provide a comprehensive overview of the use of these five types of economic instruments of environmental policy.

4.2.4 Rationales for employing economic and financial incentives

Before describing the present use of economic instruments and putting it into a longitudinal perspective (Section 4.3), it is perhaps appropriate to assess the current state of the art of economic instruments (based on Opschoor and Turner, 1994).

If, due to market failure and/or government failure, the economic process will not automatically move the economy toward a socially optimal allocation, then the question arises as to how to address this. Standard economic analysis has suggested two basic routes: (i) private negotiations and legal action, or (ii) policy interventions aiming directly at an alteration of market prices (e.g., through charges), or redefining and altering the structure of property rights (e.g., by licensing, zoning, standards, etc.). On route (i), Coase (1960) has suggested that a bargaining approach might suffice in reaching a social optimum, although there are many reasons (Pearce and Turner, 1990; Barde, 1992) why bargains do not, and cannot, easily occur. In their absence, there is a case for route (ii), government intervention. Standard economic analysis suggests particular types of intervention, including so-called economic instruments (i.e., Pigovian charges or Baumol–Oates charges; trading in emission rights or quotas). Economic theory also provides a framework within which one could theoretically compare and evaluate a number of (first- and second-best) instruments from the viewpoints inherent in economic analysis: allocative efficiency (including cost-effectiveness) and dynamic efficiency. Charges tend to be a lower-cost method of achieving a given standard or emission level, than a uniform standards policy if marginal abatement costs differ between sources. Moreover, charges are often alleged to provide stronger incentives to firms to identify and even develop clean technologies with lower marginal abatement costs, since the financial reward for doing so exceeds the benefits of a standards-based approach (at least in cases of rising marginal abatement costs). In a dynamic setting, different environmental policy instruments have different impacts on both the degree of technological innovation and the process of diffusion (or penetration) of new technologies; charges are often presumed to be capable of mobilizing technology pull and push forces relatively effectively. Marketable permits offer promises of efficiency similar to those of charges. By giving the polluters a chance to trade their pollution emission/discharge permits, the total cost of pollution abatement down to some predetermined acceptable level, is minimized. As long as polluters have different costs of abatement there is an automatic market – low-cost polluters will sell permits and high-cost polluters will buy them. Trading holds the promise of a cost-effective total abatement result, as we shall see in Section 4.3.

The Instruments Debate

Until fairly recently, direct regulation was the almost exclusive strategy to address environmental issues. Several developments since the late 1960s or early 1970s may help explain why the subject of instrument choice has gained increasing prominence on the public agenda reported above (Opschoor and Vos, 1989). First, increasing (and empirically based) doubts emerged as to the effectiveness and efficiency of (further) regulation: increasing enforcement problems and doubts about compliance with direct regulation led to a tendency to favor "deregulation". Second, the budgetary constraints faced by most governments led to an interest in instruments that promised efficiency and "built-in" compliance, and a capacity to generate funding. The use of environmental considerations as a rationale for imposing new charges proved to add to the legitimacy (at least in the eyes of large sections of the public) of what effectively boiled down to new taxation. Finally, a growing belief in the beneficial properties associated with harnessing market forces emerged.

Widespread and general as these tendencies may have been, the actual development of environmental policy instruments has followed different courses in different parts of the OECD region. This may be due to the diversity of political and administrative contexts that prevail in OECD member countries; such differences might give rise to a divergence in initial instrument preferences, that might subsequently be reinforced by the dynamics of policy decision making.

As far as differences in the structure of contexts is concerned, one could refer to a number of general features such as: (i) general political outlooks, e.g., on market intervention; (ii) the political configurations in which countries operate (e.g., the EC, EFTA/Nordic Council, NAFTA); (iii) national political structures (federalism, etc.); and (iv) administrative cultures and societal responses to (particular types of) intervention. But there may also be differences in features that relate, in a more direct way, to environmental policies, such as: (i) the basic tenets of environmental policy (e.g., quality- or source-oriented); (ii) priorities attached to environmental problems and public support for environmental policies; and (iii) distribution of responsibilities for economic sectors (and environmental compartments) over ministries, policy levels and agencies.

In all of these features, there are differences between the OECD member countries, and even between subregions such as the European Union, the North American Free Trade Agreement (NAFTA), etc. Hence, one would expect a range of different approaches. However, there appears to have been a convergence on the basic tenets of environmental policy at a fairly early stage. This initial convergence manifested itself in an almost "natural" development of a command-and-control approach for environmental policy out of its predecessors, such as policies in public health and water management, and in the early adoption by the OECD (1972) of the "polluter-pays" principle. Of course, as in any other policy field, there is periodic performance review and reconsideration; in such circumstances and given the uncertainties surrounding the costs and benefits of specific options in terms of policy and instrument development, the other features may become important in understanding the divergences in instrument development that may occur. As we shall see, in the USA there has been a relatively long tradition in applying trading approaches to environmental problems in an attempt to put more flexibility into an otherwise rather rigid, command-and-control approach. Elsewhere (as in Europe) the taxation or charging approach was considered or even used from the early 1970s onward, with differences in charge bases between the EC (typically on effluents) and Scandinavia (typically on products). There could even be some rigidity in the patterns of development of sets of policy instruments that could be accounted for by a view of decision making as one driven by rationality bounded by severe constraints in terms of awareness of the availability of options and their impacts, leading to a preference for options within known avenues. Such "satisficing" approaches may have been behind the fact that, as we shall see, in the USA most advances have been made in the development of trading approaches, whereas in Europe there appears to be a preference for the taxation approach.

The fairly widespread preference for a regulatory approach as the foundation of a system of environmental policy instruments that prevailed nevertheless, has been an important feature. In fact, this is not as strange as it may look from a purely economic perspective: it is understandable at least to a very large degree from the point of view of the political acceptability of instrument sets. One can see, the process of choosing environmental policy instruments is influenced by the motives and interests of several important groups of actors: government, industry, and environmentalist organizations. Assuming that these interest groups have a "rent-seeking" behavior in a very broad sense, one may expect support for regulatory approaches at least from the sides of industry and governments as well as environmentalist organizations, possibly at the expense of consumers (Verbruggen, 1994). In fact, the very idea that choices between charges and standards, for example, are made in a setting of seeking to achieve objectives by selecting the most efficient instruments, may be misleading. Rather, in actual political processes decision makers may be faced with series of very complex "packages" of objectives, instruments, side-payments or other compensatory or mitigating measures, etc., over very wide ranges of policy matters, and the effectiveness and efficiency issues that environmentalists and economists always assume to be predominant, may in reality be relatively insignificant. Policy analysis shows that there may be grounds for assuming that the policy arenas in which instruments for water and air quality policies were shaped, might not be conducive to incentive-based approaches, but might rather favor regulatory measures, since these are less likely to lead to unsettling conflicts between the parties involved (de Savornin Lohman, 1994). The basis for regulatory control is some form of legislation. Compliance is mandatory and often sanctions for noncompliance exist. While the standards approach is biased against technological innovation (i.e., it provides no direct incentive for regulated polluters to exceed their prescribed target level of abatement) it still proves to be relatively attractive to control agencies. The reason for this is that it provides a measure of "certainty" of the policy result in terms of environmental effectiveness (as long as there is adequate monitoring/ and enforcement). This certainty (which is the most obvious advantage of direct regulation; Opschoor and Vos, 1989) is particularly important when persistent and toxic substances are being released into the ambient environment (Opschoor and Pearce, 1991). Other advantages of direct regulation have been listed by Bohm and Russell (1985).

A number of new policy tendencies have emerged that could have a positive impact on the future role of economic instruments. First, one may observe an increased tendency toward reduced government intervention (deregulation) in the wake of the changes in central and eastern Europe at the end of the 1980s; this could lead to a more prominent role for economic instruments, particularly charges. Charges may provide incentives to change polluters' behavior in a more generic way, and may raise revenues for (self-)financing of environmental measures. Second, there are tendencies toward policy integration that could induce harmonization of economic instruments applied in different policy sectors, and the development of new, broad-based economic instruments; this is apparent from the results of UNCED and the finally concluded Uruguay Round of the GATT (and the intentions of further work on trade and the environment). Finally, in the industrialized market economies there has been a shift of attention from curative to preventive policies, which could result in a stronger role for instruments such as product charges and deposit–refund systems. This tendency is associated with (and is at least partly explained by) the increase in both the investments in and the costs of environmental protection in relation to GDP (the costs have typically increased from less than 1% in the first decades of environmental policy, to a current 2%, and beyond in the decades ahead). Economic efficiency will increasingly become a condition of social acceptability of environmental and sustainable development policies.

Hence, despite the justifications for direct regulation, one could expect that economic instruments will gain support and recognition. If this is valid, then one might see the emergence of more economic instruments, perhaps even already in recent years, but more certainly in the years to come. Some of the considerations of group interests and rigidities discussed above may favor "policies-as-usual" approaches in the instrument choices that lie ahead, but there are also strong forces in favor of convergence within the set of economic instruments.

4.3 Analysis of Developments in OECD Countries

This section begins by presenting some aggregated results of the 1992 survey of the use of economic instruments in OECD countries (Vos *et al.* 1994), with a focus on air pollution. The survey registered policy instruments that were in effect on 1 January 1992, in 23 OECD countries, and also included "proposals", i.e., instruments scheduled to be in operation on or before

1 January 1993. Section 4.3.2 compares these results with those of the 1987 survey (Opschoor and Vos, 1989) and analyzes the changes. Section 4.3.3 discusses some implications of this analysis.

4.3.1 The use of economic instruments in OECD countries, 1993

Table 4.1 provides a summary of the economic instruments used in the OECD countries in 1992. For Australia, Belgium, Canada, and the USA, instruments applied at the nonfederal level have been included. The table should be interpreted with care, as "policy instruments" within one instrument type are simply added, while in fact their relative importance differs a great deal. For instance, the US Emissions Trading Program, covering tens of thousands of sources, involving hundreds of trades and consisting of various subprograms (bubbles, netting, offsets, banking), counts as "one" tradable permit system, whereas the "plant renewal clause" in the German air pollution legislation is essentially a netting provision and has hardly been implemented so far.

There are some quantitative and qualitative differences between countries to which I would like to draw attention. First, while countries at the top of *Table 4.1* generally do have ambitious environmental policies, it is not true, judging from the data for Germany, Switzerland and Japan, that conversely, all countries with ambitious environmental policies apply economic instruments extensively. In other words, economic instruments, even if broadly defined to include revenue-raising instruments, have not yet permeated on a substantial scale into the environmental policies in all environmentally aware countries. A second observation relates to the nature of the instruments applied in various regions. When account is taken of the number of nonfederal instruments included in the data for Australia, Canada, and the USA, *Table 4.1* shows that the Scandinavian countries are leading in the application of economic instruments, especially charges. The USA is ahead in the application of tradable permit systems. This suggests that differences in the administrative or political contexts in which instruments are to be applied, may have played a role in shaping the development of the instrument mixes currently in use in the environmental policies of various countries and regions. Third, countries appear to differ when the financial or economic significance of economic instruments for environmental policy are considered; this difference will be highlighted below.

One would like very much to be able to determine the economic significance of the instruments used in environmental policy, especially charges.

Table 4.1. Economic instruments used by OECD countries, January 1992.

	Charges on emissions[a]	Charges on products[b]	Deposit–refund schemes	Tradable permits
USA	5 (2)	6 (1)	4	42
Sweden	3 (2)	11 (2)	4	2
Canada	3 (2)	7 (3)	1	22
Denmark	3 (2)	10 (2)	2	
Finland	3 (2)	10 (2)	2	
Norway	4 (2)	8 (2)	3	
Australia	5 (2)	1 (0)	3	12
Netherlands	5 (2)	4 (2)	2	
Austria	3 (1)	4 (2)	3	
Germany	5 (2)	3 (3)	2	1
Belgium	7 (2)	2 (2)	1	
France	5 (2)	2 (1)		
Switzerland	3 (2)	2 (2)	1	
Italy	3 (2)	2 (0)		
Iceland	1 (1)	1 (1)	2	
Japan	3 (1)	1 (1)		
Portugal	2 (0)	1 (1)	1	
Ireland	2 (2)	1 (1)		
Greece		2 (1)	1	
Spain	3 (2)			
UK	1 (1)	1 (1)		
New Zealand	1 (1)			
Turkey			1	

[a]Data in parentheses are user charges.
[b]Data in parentheses are tax differentiations.
Source: Vos *et al*, 1994.

Unfortunately, consistent data on charge revenues are not available from the surveys, but based on what information they did yield and on OECD tax revenue statistics, some sense of the order of magnitude can be obtained (Vos *et al.*, 1994). Most charges have been earmarked as revenue; for some charges the way revenues are spent is unknown; in a small number of cases revenues from charges feed the general budget. In Sweden the revenue from environmental charges (excluding user charges) in 1991 was about 0.75% of GDP. The CO_2 charge accounted for 0.65% alone. In fact, the Swedish CO_2 charge is an integral part of a system of energy taxation with substantial revenue; but general energy taxes were not included in this survey. In Denmark, environmental charges (excluding user charges and apart from the carbon tax) account for approximately 0.15% of GDP.

Norwegian revenue from environmental charges (apart from user charges) is 0.75% of GDP. In the Netherlands the revenue from environmental charges, including the water pollution charge, is 0.6% of GDP. The water pollution charge alone accounts for 0.4%. In Finland, revenue from environmental charges, including user charges, was 0.9% of GDP in 1992; excluding user charges, the revenue was 0.4% of GDP. In 1991 the revenue from the CFC and the hazardous waste charges in the USA (Superfund) amounted to 0.3% of GDP.

Incentive Intentions and Impacts

The environmental effectiveness of a policy intervention may be expected to be significant especially when the intervention is done in the context of the second basic strategy mentioned in Section 4.2.1: to influence the decision making process at the micro-level. This is what is normally meant by "incentive impact"; such an intervention can then be labeled "incentive-based". An attempt was made to establish whether a specific scheme had been intended as an incentive based instrument (e.g., by looking at the charge rate, or from intentions as expressed by policy makers in policy documents), in order to distinguish the truly economic instruments from revenue-raising ones. Basically, one should distinguish the level of the intentions when the instrument was established, and the level of effective impacts of the instrument: they can be intended and/or shown to have incentive impacts, or not. Thus, a two-by-two matrix of possible cases emerges (Opschoor and Vos, 1989). A hypothesis worth testing would be that economic instruments with incentives built in would have emerged more prominently since the early days of environmental policy, now that economic analyses have penetrated further into environmental policy, and the reasons for taking an efficiency-oriented approach have become so much more evident.

Table 4.2 summarizes the evidence on incentive intentions and impacts in the case of emission and product charges. We take it that in the case of deposit–refund schemes the incentive impact is obvious and is also shown to have worked in economic reality. It must also be noted that the table leaves out all tax differentiation schemes on automobiles and gasoline that actually do have incentive intentions and very often incentive impacts as well. The work underlying the table may have been biased against imputing the label "incentive" to the instruments considered: actual incentive effects were recorded if either formal research has demonstrated incentive impacts or if environmental impacts have been achieved that cannot be accounted for by other factors. In reporting on actual incentive effects, the "unclear"

Table 4.2. Incentive effects of charges and taxes.

	Emission charges and taxes: incentive intended (actual)	Product charges (excl. tax differentiation on automobiles and gasoline): incentive intended (actual)
Yes	15 (3)	21 (7)
Data inconclusive	8 (7)	7 (10)
No	11 (-)	23 (20)
No data available	- (24)	- (14)

NB: All tax differentiation schemes on automobiles and gasoline (30 schemes) can be said to at least have incentive intentions.
Source: Adapted from Vos *et al.*, 1994.

category applies to cases where there is some, but inconclusive, evidence of incentive effects. For instance, if a charge is applied in conjunction with other policy instruments, it is difficult to identify separately the effects of the charge and the accompanying policies. Hence, the "unclear" – and "no data" – categories are likely to contain a number of instruments with incentive impacts. Nevertheless, it is fair to state that the number of cases of charges with manifest incentive impacts is lower than expected, especially with emission charges. For both emission charges and product charges, the intentions were much more oriented toward an incentive impact, as empirical reality convincingly corroborates. One more positive way of interpreting the table is that possibly it takes more time and especially research for these incentive impacts to manifest themselves in the economic process and to be picked up through empirical analysis. Another way of interpreting it might be that very often even when incentive intentions are present, the intensities with which the instruments are applied (e.g., the charge rates) are such that incentive impacts should be expected to be negligible or at least very hard to detect: charge rates very often are too low. The available evidence on water effluent charges, for example, indicates that the level of the charge does indeed have an impact in terms of environmental performance of the agents concerned (Opschoor and Vos, 1989).

There is less ambiguity about the incentive impacts of other types of economic instruments. All deposit–refund systems reported can be assumed to have incentive intentions. Deposit–refund schemes (mainly in the area of beverage packaging) show return percentages varying between 40 and 100%, with an average near 80%. The US Emissions Trading Program has achieved cost savings, but the number of trades has remained limited and most trades have been within companies. The limited effects are due to the restricted scope of the program (allowing only trading of emission reductions beyond regulatory baselines), the trading rules (promoting primarily intra-company trading), and the uncertainty created by the joint application of discretionary command-and-control regulation alongside the tradable permit program (Atkinson and Tietenberg, 1991). The Acid Rain Control Program, which began in 1993, has more characteristics of a full-blown tradable permit program. Enforcement incentives include performance bonds and penalty fees, the rates of which are not fixed, but depend on the damage inflicted or noncompliance benefits reaped. They clearly have incentive intentions. Data on incentive effects are not available.

In conclusion, incentive impacts were intended in most schemes considered here, even with many charging schemes (although one also finds many examples of revenue-raising instruments rather than incentive-based ones). However, whether an instrument actually has incentive effects is another matter: this may also depend on the intensity with which the instrument is applied. Looking at charges in particular, it is noteworthy that even when incentive impacts were present, the incentive effects are not always manifest. This is mostly due to the failures to obtain empirical evidence: data are (as yet) inconclusive or insufficiently available. Other problems in establishing the significance of an instrument in terms of its incentive impacts are: (i) very often there is no clear baseline or (empirical or simulated) reference with which to compare the impact of the introduction of a particular instrument; and (ii) many instruments are components of more complex packages, and attributing the environmental effectiveness of these packages to their specific components is not very easy. More research is urgently needed.

Air Pollution-Related Economic Instruments

Specific air pollution-related instruments exist in the categories of charges (both emission and product charges), tradable permits and enforcement incentives. Details of the mechanisms upon which these instruments rely are not provided here (see Vos *et al.*, 1994; and other contributions to this volume); rather the emphasis will be on some aspects of their performance in terms of efficiency. *Table 4.3* presents an overview of the air pollution charges as covered by the OECD survey.

Some comments are appropriate here on the incentive impacts of these emission charges. As to the Canadian charge, an incentive effect is conceivable since permits are revised if applicants can demonstrate that actual emissions are below permitted levels; there are as yet no data on actual

Table 4.3. Air pollution charges in six OECD countries.

		Incentive		Revenue
Country	Charge base and rate	Intended	Actual	spending
Canada[a]	Permit fee for air pollution	–/+	..	Air quality control
France	Acidifying emissions: ECU 19.0/ton (SO_2, NO_x, H_2S, N_2O, HCl)	–	..	Subsidies
Japan	SO_x emissions: rates differ regionally, between ECU 0.5–4.5/Nm^3	+	#	Compensation of health damage
Portugal	SO_2, NO_x	..	..	Air quality control
Sweden	NO_x-emissions of energy producers: ECU 4.7/kg NO_2	+	+	Rebated to energy producers
(USA)	Criteria pollutants, $\geq$ ECU 16.4/ton	–/+	..	Air quality control

Symbols: + = yes, – = no, .. = no data available, # = unclear.
[a]The instrument came into force in January 1993.
Source: Vos *et al.*, 1994.

incentive effects. In Japan ambient concentrations of SO_2 have been substantially reduced, but it is unclear to what extent this has been due to the charge. Anyhow, some incentive impact could perhaps be expected from the charge. The incentive effect of the Swedish NO_x charge has surpassed all expectations. The charge speeded up considerably compliance with the sharper emissions standards that will be imposed in 1995. The US air pollution permit fees, imposed on 1 January 1993, are also expected to have incentive effects.

Apart from emission charges, air pollution abatement frequently makes use of product charges levied on products (e.g., mineral oils) or on product characteristics (e.g., carbon or sulfur content). The so-called "tax differentiations", such as those on leaded and unleaded gasoline, are normally included in this category. Tax differentiation is typically applied in relation to vehicles and road transport. Some 11 countries have reported vehicle sales tax differentiation schemes based on weight, fuel efficiency, compliance with standards, or the presence of catalytic converters. In many countries there are regulations specifying that catalytic converters will be compulsory at some future date, so the aim of the tax differentiation has been to speed up the market penetration of vehicles equipped with catalytic converters.

Table 4.4. Rates of carbon taxes in six OECD countries.

Country	Rate per ton CO_2	Implicit rate of existing excises	Incentive[a] Intended	Actual
Denmark	ECU 5.5–11.1	ECU 26.3	+	..
Finland	ECU 1.1 for non-traffic fuels	ECU 19.1	–	..
(Italy)	ECU 1.7	ECU 39.9	..	..
Netherlands	ECU 0.4	ECU 15.9	–	..
Norway	ECU 13.8 on coal and coke ECU 15.7 on fuel oils ECU 40.6 on gasoline ECU 40.6 on natural gas	ECU 32.5	+	..
Sweden	ECU 37.9	ECU 38.2	+	#

[a]For explanation of symbols, see *Table 4.3.*
Source: Hoeller and Coppel, 1992.

Differentiation of the annual vehicle tax between cars with high and low emissions is a similar instrument. Such systems exist (or will soon be introduced) in four countries; no data are available as yet on their incentive impacts. Furthermore, all OECD countries levy excises on gasoline. In some countries a surtax on leaded petrol is part of the gasoline excise. Revenues flow into the general budget, except in Iceland where it is earmarked for road construction and maintenance. Market shares of unleaded petrol have risen, but this may also be due to the increasing number of vehicles equipped with catalytic converters and regulations on this issue.

A special set of product charges related to air pollution is charges on fossil fuels, notably carbon taxes and sulfur taxes. Carbon taxes are levied on top of existing excises on fossil fuels. *Table 4.4* provides the rates of explicit carbon taxes, and also the carbon tax rates implicit in existing excises (cf. Hoeller and Coppel, 1992). Charges in Finland and the Netherlands are too low to have incentive effects. In Denmark, Norway and Sweden the carbon charges have been instituted only recently so that the survey was unable to capture any incentive impacts. In evaluating incentive effects, the impacts of differences between countries in existing excises on fossil fuels should not be lost sight of. In Sweden energy and CO_2 taxation is so high that some district heating plants are changing from fossil fuels to biofuels. Norway, Sweden, and Finland also have taxes on the sulfur content of fuels.

The Swedish scheme has been officially evaluated and appears to be highly effective. The charge is repayable if a taxpayer can demonstrate an actual reduction in sulfur emissions.

Under air pollution-related product taxes, some reference should be made to energy taxes, even though in the past they may have been introduced for anything but environmental reasons. Energy taxes might be considered at least in practice to have had substantial significance for air pollution policy, in that they have affected energy consumption and thereby (indirectly or directly) emissions of air pollutants. Section 4.3.2 discusses the new generation of energy taxes, that is energy taxation with explicit reference to environmental considerations (e.g., ecotaxation, or taxation of energy in relation to climate or global change issues).

Finally, ozone-depleting chemicals are taxed in four countries. In Australia, some states have imposed fees on CFCs in addition to the national charge. The Danish and Australian charges are levied on domestic production and imports. The Danish charge is administered alongside a set of regulatory controls; Danish CFC consumption fell by 60% between 1986 and 1991. In the USA a charge is levied on all ozone-depleting chemicals. The charge is applied as a complement to a system of tradable production quotas, and is designed to skim off windfall profits created by production limitations imposed by the tradable quota scheme. It will be raised progressively in the coming years. The US charge has had incentive effects on the use of CFCs in soft foams.

We turn now to air pollution-related economic instruments other than charges and taxes. This brings us especially to the trading schemes for pollution. By way of an introductory observation, it is to be noted that these instruments very transparently show a feature that is much more widely applicable: the fact that very often economic instruments are used in combination with direct regulations. In the case of emissions or quota trading, the overall levels of pollution (or reductions thereof) are exogenously determined by public authorities; within this overall level, trade is encouraged as an instrument to achieve cost-efficiency. Germany, Canada and, notably, the USA have tradable permit systems related to air pollution. In the USA the aim is to reduce production and consumption of ozone-depleting chemicals (CFCs and halons) by 20% in 1993 and 50% in 1998 (baseline 1986). Production rights have been grandfathered to existing producers on the basis of their 1986 production levels. Consumption rights have been allocated to producers also (as production minus exports) and to importing firms. Producers' increases in their holdings are constrained as well. To skim off producer rents created by the mandated production cutback a charge has been instituted. Another case in point is that of the US Acid Rain Allowance Trading program, under which almost all emissions rights are grandfathered on the basis of actual emissions in 1985 and average fuel usage in 1985–1987; about 3% are sold in open auction for a pre-fixed price. The USA also has two enforcement incentives, one of which is a noncompliance fee for SO_2 emissions (instituted in January 1993), as a complement to the scheme of tradable allowances that has been instituted in the revised Clean Air Act for acid rain control. The fee is approximately 300% of the expected market price for allowances. Much of the significance of these trading schemes has to do with their demonstrated (as in the case of the USA's old trading program; see, e.g., Hahn and Hester, 1989) or expected cost-effectiveness. For instance, the new RECLAIM scheme for NO_x and SO_x abatement (with reductions of up to 75% in 2003) introduced in the Los Angeles area, is expected to give rise to efficiency gains of approximately 35% of overall additional abatement costs.

4.3.2 Changes in the use of economic instruments since 1987

If we use a loose definition of economic instruments, i.e., including financial and fiscal instruments that may not have had the intention of modifying the behavior of polluters and resource users, then such instruments have been around for a long time. In a report to the OECD, Opschoor and Vos (1989) reviewed the situation in the OECD countries in 1987, and identified a total of 150 economic instruments (or over 10 per country surveyed), some of which had been in use for decades. Excluding a number of instruments that are beyond the scope of the present survey – subsidies, administrative charges, and liability – still gives a total of nearly 100, or almost seven per country surveyed. Of the 150 instruments identified, about 80 were charges, about 40 were subsidies, and the remainder were other types such as deposit–refund systems and trading schemes. Looking at the various economic instruments in terms of their original purpose and how they actually performed (in both cases, either providing an incentive or raising revenue), one finds that in terms of the numbers involved, less than half of the economic instruments reviewed had the intention of generating an economic incentive, and over half were intended to raise revenue. Only one-third may have effectively had some incentive impact. Thus, in 1987 environmental policies in the OECD countries were command-and-control policies with some financial and economic add-ons.

Comparing the 1993 survey results with those on the 1987 situation is not straightforward. First, more countries are covered in the present

Table 4.5. Comparison of the 1987 and 1992 survey results.

	Charges on emissions		Charges on products		Deposit–refund schemes		Tradable permits		Enforcement incentives	
	1987	1992	1987	1992	1987	1992	1987	1992	1987	1992
Finland	3	3	6	10	1	3	–	–	–	–
France	5	5	1	2	–	–	–	–	–	–
Germany	5	4	2	3	1	2	1	1	–	–
Italy	3	3	2	2	–	–	–	–	–	–
Netherlands	6	5	5	4	2	2	–	–	–	–
Norway	2	4	8	8	2	3	–	–	–	–
Sweden	3	3	7	11	3	4	–	–	2	2
USA	4	5	?	6	?	4	3	4	1	2

Source: Vos *et al.*, 1994.

survey: data from 23 countries are included, compared to 15 in the 1987 survey. Second, apart from the responses to the questionnaire, many more data sources were used in 1993, whereas the earlier survey relied on official responses and some in-depth case studies. Finally, the coverage of instruments in the present survey is more restricted in that it does not deal with subsidies, administrative charges and environmental liability. Nevertheless, a rough comparison is possible for eight countries that were surveyed in detail in both the previous and the present surveys (*Table 4.5*).

Changes have been minor in France, Germany, and Italy, moderate in the Netherlands and Norway, and extensive in Finland, Sweden, and the USA. A further assessment of the implications of *Table 4.5* is given in Section 4.3.3; here a few developments pertinent to air pollution policy are underlined. Since 1987 France has instituted, among other economic instruments, a tax differential on leaded and unleaded gasoline. In Germany a differentiation of vehicle taxes has been introduced. Norway has instituted a carbon tax. Together with the Netherlands, Norway is actively promoting international interest in economic incentives for abating greenhouse gas emissions. Finland and Sweden have actually implemented a range of new product charges, to a considerable extent with incentive purposes. Finland has instituted charges on carbon and diesel oil. In the government service there is a permanent committee for investigating and proposing environmental charges. Before 1991 Sweden introduced a differentiation of car sales taxes and a charge on domestic aviation fuel. In 1991 the Swedish tax system was reformed, involving tax revenues of approximately 6% of GDP. The reforms included a broadening of the VAT base (to include energy products, for example) and environmental charges on energy products, together with reductions in personal income taxes and the existing general tax on energy products, respectively. The new environmental charges include a carbon tax, a tax on sulfur in fuels, a tax differentiation of petroleum products, and a charge on NO_x emissions of energy producers. Since 1987 the USA has deployed the instrument of tradable emissions rights to control emissions of ozone-depleting chemicals and SO_2. A charge on ozone-depleting chemicals has been instituted, alongside the tradable consumption and production quota system. With the revision of the Clean Air Act new opportunities for economic instruments were created: from 1993 states were authorized to levy permit fees on the basis of recorded emissions. The present survey includes a number of product charges and deposit–refund schemes applied by US states.

Current National and International Initiatives

In addition to the economic instruments in use in 1992, there has been a further increase in political interest in employing these instruments. In at least eight countries official task forces are carrying out (or have recently concluded) feasibility studies, and six other countries have stated a general intention to increase the application of economic instruments (de Savornin Lohman, 1994).

At the national level, several developments not reflected by the survey reported here, must be mentioned. In Belgium, for example, a set of product charges has been implemented in 1994 (on products such as disposable beverage packaging, pesticides, paper, disposable razors, and cameras). Switzerland is considering incentive charges on volatile organic compounds, the sulfur content of heating oil, fertilizers, pesticides, and batteries. Austria is considering a charge on wastewater, fiscal incentives to reduce chlorine emissions, a deposit–refund scheme for refrigerators, and energy and carbon taxes. Denmark, Finland, and Sweden are continuously adapting their systems of fiscal incentives. In Germany there has been a long-running policy debate on economic incentives in waste management and transport. The UK is considering charges on wastewater and in the area of waste management. The recent introduction of another air pollution trading scheme in the Los Angeles area has been mentioned already.

These national developments are in line with the general interest manifested in UNCED documents such as *Agenda 21* and the Rio Declaration, but there are some more explicit and tangible indications as well. In international forums, taxes and tradable permits for greenhouse gas abatement are on the agenda. For example, the European Commission has discussed a proposal for an energy carbon tax, to be implemented conditional on the implementation of fiscal incentives by major trading partners, and in the USA the Clinton Administration has initiated a debate on energy taxation (although so far this has only generated a small increase in the gasoline tax; there may be subsequent rounds in that debate). The OECD has also paid much attention to the use of economic instruments for abating global environmental risks (OECD, 1992a, 1992b, 1993).

4.3.3 An assessment and discussion

Below we summarize, in three conclusions, the most important changes that have occurred since 1987 placing special emphasis on eight countries: the six countries that were studied in depth in 1987 – France, Germany, Italy, Netherlands, Sweden, and the USA – and Finland and Denmark, for which sufficient information is available.

First, there have been quantitative changes in the use of economic instruments. Since 1987, the use of economic instruments appears to have intensified. Comparing 1987 data with "solid" 1992 indications, the number of economic instruments used in the eight countries has grown by some 25%; including also the instruments put into operation since 1992, the number has grown by close to 50%. Looking at the totals for all OECD countries on which we have information for 1987 and 1993, the growth in the reported use of economic instruments would be even larger, but this might also reflect the greater accuracy of the 1992 survey.

Second, there have been changes in the type of instrument. Instruments that have been introduced more frequently are product charges and deposit–refund systems. In the eight countries, comparing 1992 with 1987, the number of these instruments increased by 35% and 100%, respectively, and this trend appears to be continuing. It is noteworthy that emission charges were not used more frequently in 1992, as far as we can tell, in the eight countries, although there may have been a small increase in the reported number since 1992. If we look at all OECD countries on which we have information for both 1987 and the present, then there appears to have been an increase, but again, this may also reflect the more complete survey of instruments in 1993. Deposit–refund systems still function mainly in the field of beverage packaging.

Finally, there have been changes in incentive impacts or the significance of the instruments. It is clear that too little information exists to provide firm judgments on this very important aspect. Looking at the available information (i.e., using solid data on the 1992 instruments), we find that on emission charges one can say with some confidence that an incentive impact is now at least formally intended in some 45% of the charges (whereas in some 30% it is not), but in about 90% of the cases the information on actual incentive impacts is inconclusive or unavailable. With respect to product charges, we know that incentive impacts were intended as often as they were not (45% in each case). When it comes to actual incentive impacts, there are inconclusive (or no) data on close to half of the applications, and such effects seem to be absent in at least 40%. The lack of conclusiveness of the available information has to do with the fact that very often product charges are applied in conjunction with other instruments. The US Emissions Trading Program has achieved cost savings, but the number of trades has been lower than expected; more incentive impacts might be expected from the Acid Rain Allowance Trading program. The Lead Trading Scheme has performed quite well.

Assessment

In terms of the policy tendencies presented at the end of Section 4.2.4, the role of charges has indeed been extended. Although an increasing number of charge schemes now show incentive purposes, not much evidence has been mobilized to show that these charges have actually had incentive impacts. Nor is it likely that such instruments will replace direct regulation, as part of a process toward reducing government intervention. Some examples of new incentive charges are: the CO_2 charges in Denmark, Norway, and Sweden; the Swedish NO_x charge; the Danish waste disposal charge; and a number of product charges (on fertilizers in Sweden, on ozone-depleting chemicals in the USA, and on packaging in Finland and Norway). There may, in all, be more charges now than there were some years ago, but this expansion is restricted largely to product charges. Moreover, the revenue-raising capabilities of these charges have remained their dominant function. Tax differentials on the sales prices of new cars with catalytic converters and on unleaded petrol may well have contributed to the rapid penetration of these products on the market. The extension of the number of deposit–refund systems might be a signal of intensified government attempts to shift responsibility for packaging waste to the societal parties concerned, and to reduce their own role. A remarkable development has been the introduction of deposit–refund systems for plastic bottles. These were hardly mentioned in the 1987 survey, whereas the present inventory shows widespread application in at least 11 countries. On the second tendency discussed in Section 4.2.4

(i.e., policy integration) there have also been some developments (Vos *et al.*, 1994), but these are hardly reflected in the role of economic instruments. The wider attention paid to carbon taxes may be regarded as an example, since it is an instrument of both environmental policy and of energy policy. Although curative policy is still a major policy issue, preventive policies have increased in importance. The fact that more product charges and deposit–refund systems are in operation now than in 1987 may underline this tendency.

Discussion: Instruments

There have indeed been qualitative and quantitative developments in the use of economic instruments, in line with expectations based on some overall policy tendencies. Yet, given the theoretical as well as political arguments put forward in Section 4.2.4 in favor of economic instruments, the rate of implementing incentive instruments appears to be rather slow. One explanation for this might be that much of the original debate on economic instruments has remained too remote from the realities of the economic process and the policy arena; a more pragmatic approach would modify the rather sweeping generalizations and expectations to more realistic proportions. Discussions on the use of economic or other instruments (or mixes thereof) are increasingly stressing the need for a more pragmatic approach. Partly as a result of empirical studies, there is now much less dogmatism or rhetoric in the dialogue on environmental policy instruments; this pragmatic approach is indeed beginning to prevail. There are fewer categorical pronouncements in favor of or against certain types of instruments; rather, the interest is in providing realistic assessments of the pros and cons of different mixes of instruments in the specific policy and application contexts within which they are to operate. Some of the main criticisms in the more empirically oriented literature on the theoretical argumentation in favor of economic instruments have been:

1. Economic analyses do not always convincingly reflect economic realities. Real markets do not always work as theory assumes (because there is very imperfect information, there are rigidities and/or transaction costs, there are too few parties in the market, etc.). This criticism has some merit, as empirically based doubts *vis-à-vis* the alleged effectiveness and dynamic efficiency suggest. Looking at the aspect of effectiveness, it must be realized that financial and economic instruments provide an incentive, but the significance of the incentive depends on the elasticities that operate on behavior (price elasticities, substitution elasticities,

income elasticities), on the strength of the signal given (e.g., the level of the charge), and on the availability of substitutes or alternative actions. Sometimes, high charges alone will not have sufficient behavioral significance and hence their environmental effectiveness is low, or doubtful. Most charges introduced so far in OECD member countries have proven to be too low to have had an incentive impact, even if this was intended. On the other hand, evidence on the impacts of some charges and of energy price fluctuations suggest there is scope for effective pricing policies and charges, etc. On the aspect of dynamic efficiency, some qualifications are also called for. In the long run, this aspect is much more important than allocative efficiency, since it relates to the development of new innovation trajectories and the shaping of new options in consumption, production, and investment. Until now, the impacts of the various policy approaches on actual patterns of innovation and diffusion of environmentally friendlier technologies are still a matter for empirical analysis and debate (Georg and Jorgensen, 1990, p. 6; Kemp *et al.*, 1994).

2. Noneconomic instruments may perform equally well or even better than economic incentives, especially from the point of view of effectiveness. Looking back at the arguments in favor of direct regulation (Section 4.2.5) and at what we have observed in relation to the development of economic instruments so far, one may conclude that noneconomic instruments do have a strong *raison d'être*. In fact, they still dominate the instruments selected, and very often economic instruments are used merely as adjuncts to regulatory approaches. The performance of "mixes" of instruments, in terms of efficiency and effectiveness, is as good as (if not actually better than) that of "pure" instruments, especially the purely economic ones, and there are good reasons for this. In other words, a shift from a regulatory approach toward one based on economic incentives alone is unlikely to occur, nor would it be desirable. Combinations of instruments are preferable, and within such "cocktails" economic incentives will play a crucial role.
3. Economic instruments cannot be as efficient as was claimed initially by economists, since the analysis bypasses a number of issues related to implementation and enforcement. There is growing recognition of the diversity of application contexts and the relevance of this diversity for instrument selection. The effects of policy instruments depend on the economic, political, and administrative contexts in which they are applied. The complexities of the interactions between environmental and economic processes, as well as the dynamics of innovation, preclude

straightforward and simple broad-brush recommendations as to which instruments to use. These issues are gaining increasing attention in the literature, and there are promising examples of cross-fertilization between economists and noneconomists in collaborative efforts to address them.

4. Economic analyses and recommendations often ignore or play down realities typical of the political "arena" in which (environmental) policy is shaped in reality. One aspect of this is the issue of the distributional implications of environmental policy and instrument choice. This issue was touched upon in Section 4.2.4.

Discussion: Policy Interest

Some recent manifestations of the growing policy interest have been: (i) the Brundtland Report (WCED, 1987) and the UNCED process with its concern over how to merge environment and economics in decision making (UNCED, 1992); and (ii) the increased tendency to rely on market-based approaches, not only in OECD countries but also in economies in transition and many developing countries. In addition, environmental policy attention has shifted. Environmental policy since the end of the 1980s has become more interested in tackling environmental problems at higher spatial levels, at the continental and even global scale. At that level there is much less scope for a command-and-control approach and hence economic instruments may provide more attractive options. However, it is only fair to say that so far the typical approach to coordination has been to adopt some time schedule for achieving fixed abatement targets in terms of percentage reductions to be applied to individual countries' emissions in some reference year. Apart from its concern with international problems, environmental policy has also become more interested in diffuse and mobile sources of pollution: again types of sources for which some economic instruments (especially product charges and deposit–refund systems) could prove effective.

In current discussions on economic instruments some new elements have emerged, or some specific aspects are now receiving much more attention than they did in 1987. One of these is explicit reference to the link between institutional failure and environmental externalities. Environmental degradation has its roots in the economic process (notably its relation to types and levels of economic activities, and their development), as well as in the way decision making on economic activities and policies has been institutionalized. Conditions pertaining to the institutions that direct the economic process from within (i.e., the market and "market failure") are such that government interventions are needed to ensure the incorporation of environmental externalities in market decisions. Also, the need to remove old institutional failures (such as hidden or explicit subsidies) is now more broadly recognized. A second new element is the awareness of the risk of failure accumulation. Existing regulations and policy interventions may create a setting in which efficiency-based consideration of environmental policy instruments would at best indicate second-best solutions. Internalizing externalities via economic instruments may not make much sense in a world of seriously distorted markets (i.e., in a complex situation involving multiple market and government failures); in fact, the overall result may be ambiguous. A third new element is the growing interest in harmonization. There is a need to at least consider the harmonization of individual countries' approaches to the choice of instruments; if international trade and investment patterns are likely to be affected. Finally, one older consideration may still be important. There is a perhaps growing need for funds for financing environmental measures; this need may support the use of economic instruments – especially charges – which may have incentive side effects.

4.4 Conclusions and Outlook

Even though economic instruments are now more widely used in environmental policy, and even more applications are under consideration or being implemented, the most striking development since 1989 has been not the growth of the use of economic instruments, but the growth of interest in the subject of economic approaches and instruments – theoretically, empirically, and in terms of policy design. This observation gives support to the expectation that over time one may see an increased use of these instruments, if pragmatic arguments can be found in their favor. This holds for economic instruments in general, as well as for applications to air pollution abatement.

The potential policy relevance of economic instruments for environmental policy is by now well established and no longer rests on theoretical or academic arguments alone. The call for increasingly stringent environmental quality standards in the short run inevitably means rising policy response costs and therefore enhances the appeal of cost-effective market-based instruments. Economic instruments are potentially relevant in policy contexts that are based on a preventive approach. Their use is therefore likely to grow.

There are substantial differences in the intensities with which economic instruments are applied. Different sets of these instruments are being tried out in various parts of the world. One can still observe a wide range of opinions as to when and where to apply which type of instrument or mix of instruments. The effects of policy instruments depend on the economic, political, and administrative contexts in which they are applied. In so far as harmonization develops, these differences may diminish.

The relative roles of economic instruments and direct regulation might be changing, and they even might alter in favor of economic instruments, but this is certainly not yet a very clearly observable tendency, since also the use of other types of instruments may have evolved since 1987. Nowhere is there an observable change toward replacing the basic command-and-control approach with a purely economic one. Economic instruments are mostly complements to and only sometimes substitutes for other types of approaches. That is, they appear to operate mostly in combination with, or in support of, other instruments such as direct regulation.

Economic instruments are now increasingly being considered for application at the regional/global level. Global or regional energy charges or carbon charges are being discussed, and global and tradable carbon quotas, are being investigated.

Options for integrating environmental considerations into the policies of other sectors are far from having been exhausted. In fact, the OECD countries are now witnessing the beginning of such integration efforts. Economic instruments and generally instruments that bring market signals closer to their socially desirable levels, have important roles to play and this has increasingly become common insight. Environmental taxation may be one specific area in which this policy integration can take place, but the relationships to the fiscal framework as such remain to be worked out.

Acknowledgments

The author wishes to acknowledge the inputs from two anonymous referees and from the editors of this volume. Much of the original work for the OECD studies reviewed here was done by Messrs. Vos and de Savornin Lohman, and they are gratefully acknowledged.

References

Atkinson, S., and Tietenberg, T., 1991, Market Failure in Incentive Based Regulation: The Case of Emissions Trading, *Journal of Environmental Economics and Management* **21**:17–31.

Baumol, W.J., and Oates, W.E., 1975, *The Theory of Environmental Policy*, Prentice-Hall, Englewood Cliffs, NJ, USA.

Barde, J.-Ph., 1992, *Economie et Politique de l'Environnement*, Presses Universitaires de France, Paris, France.

Bohm, P., and Russell, C.S., 1985, Alternative Policy Instruments, in A.V. Kneese *et al.*, eds, *Handbook of Natural Resource and Energy Economics*, Vol. 1, North Holland, Amsterdam, The Netherlands.

Coase, H., 1960, The Problem of Social Cost, *Journal of Law and Economics* **3**(1):1–44.

Georg, S., and Jorgenson, U., 1990, Clean Technology: Innovation and Environmental Regulation, Paper presented at the Conference of the European Association of Environmental and Resource Economists, April, Venice, Italy (unpublished).

Hahn, R.W., and Hester, G.L., 1989, Economic Prescriptions for Environmental Problems, *Journal of Economic Perspectives*, Spring:95–114.

Hoeller, P., and Coppel, J., 1992, *Energy Taxation and Price Distortions in Fossil Fuel Markets: Some Implications for Climate Change Policy*, OECD Economics and Statistics Department, Working Paper No. 110, OECD, Paris, France.

Kemp, R., Olsthoorn, A., Oosterhuis, F., and Verbruggen, H., 1994, Policy Instruments to Stimulate Cleaner Technologies, pp. 275–301 in H. Opschoor and K. Turner, eds., *Economic Incentives and Environmental Policies: Principles and Practice*, Kluwer Academic Publishers, Dordrecht, The Netherlands.

OECD, 1991, Recommendations of the Council on the Use of Economic Instruments, in *OECD Environmental Policy: How to Apply Economic Instruments*, Organisation for Economic Co-operation and Development, Paris, France.

OECD, 1992a, *Climate Change: Designing a Tradable Permit System*, Organisation for Economic Co-operation and Development, Paris, France.

OECD, 1992b, *Climate Change: Designing a Practical Tax System*, Organisation for Economic Co-operation and Development, Paris, France.

OECD, 1993, *International Economic Instruments and Climate Change*, Organisation for Economic Co-operation and Development, Paris, France.

Opschoor, J.B., 1990, Economic Instruments for Sustainable Development, in *The Conference Report*, Conference on Sustainable Development, Science and Policy, NAVF, Bergen, Norway.

Opschoor, J.B., and Pearce, D.W., 1991, *Persistent Pollutants*, Kluwer Academic Publishers, Dordrecht, The Netherlands.

Opschoor, J.B., and Turner, K., 1994, *Economic Incentives and Environmental Policies: Principles and Practice*, Kluwer Academic Publishers, Dordrecht, The Netherlands.

Opschoor, J.B., and Vos, H.B., 1989, *Economic Instruments for Environmental Protection*, OECD, Paris, France.

Pearce, D.W., and Turner, R.K., 1990, *Economics of Natural Resources and the Environment*, Harvester Wheatsheaf, New York/London.

Savornin Lohman, A. de, 1994, Economic Incentives in Environmental Policy: Why are they White Ravens? in J.B. Opschoor and K. Turner, eds., *Economic Incentives and Environmental Policies: Principles and Practice*, Kluwer Academic Publishers, Dordrecht, The Netherlands.

Verbruggen, H., 1994, Environmental Policy Failures and Environmental Policy Levels, in J.B. Opschoor and K. Turner, eds., *Economic Incentives and Environmental Policies: Principles and Practice*, Kluwer Academic Publishers, Dordrecht, The Netherlands.

Vos, J.B., Savornin Lohman, A. de, and Opschoor, J.B., 1994, *Integrating Environment and Economics: The Role of Economic Instruments*, OECD, Paris, France.

UNCED, 1992, *Agenda 21*, UN, Geneva, Switzerland.

World Commission on Environment and Development (WCED), 1987, *Our Common Future*, Oxford University Press, Oxford, UK.

Chapter 5

Economic Instruments for Air Pollution Control in Sweden

Kerstin Lövgren
Swedish Environmental Protection Agency
Solna, Sweden

Abstract

In recent years, traditional forms of regulation have been combined with economic instruments in Swedish air pollution control policy. Major economic incentives are provided by the sulfur tax, the nitrogen oxides charge, and the environmental classification of diesel fuels. Both economic instruments and administrative regulations have played a part in reducing emissions, but the new economic instruments must be considered to have been the major driving force behind the reductions. The average sulfur content of heavy fuel oil has decreased to approximately 0.4% while that of almost all gas oil is now less than 0.1%. Total nitrogen oxides emissions from boilers liable to the nitrogen oxides charge have decreased by about 40%. The tax on diesel fuels is differentiated to stimulate the use of environmentally better grades of fuel, which are taxed at a lower rate than the standard grade. As a result, diesel fuels with better specifications than the standard grade have captured a large fraction – about 75% – of diesel sales in only a few years.

Key words: sulfur tax, nitrogen oxides charge, diesel fuel taxes, economic incentives (actual performance of).

5.1 Introduction

Economic instruments have been successfully introduced into Swedish air pollution control policy in the last few years. This chapter provides a brief overview of the new incentive mechanisms and examines the performance of three of them in some detail: the sulfur tax, the nitrogen oxides charge, and the differentiated tax on diesel fuels.

5.1.1 Economic instruments for air pollution control: An overview

The new economic instruments apply to energy installations and vehicles. In 1993 a carbon dioxide tax, a sulfur tax, and a charge on nitrogen oxides emissions were introduced, as well as systems of environmental classification of vehicles and fuels backed by tax incentives for the use of environmentally better products (see *Table 5.1*).

The introduction of the carbon dioxide tax and other environmental taxes in 1991 were part of a far-reaching reform of the Swedish tax system, one of the major aims of which was to reduce income taxes. The reduction of income taxes was financed largely by increasing indirect taxes and by broadening the base of value added tax (VAT). When the carbon dioxide tax was introduced, existing energy taxes were cut by 50%, and VAT was extended to apply to all forms of energy. Previously, only energy taxes had been levied on various forms of energy. The combined effect of these changes was a substantial rise in the overall level of fossil fuel taxes. By virtue of the reform, a total of approximately SEK 18 billion was transferred from income tax to various types of energy and environmental taxes (Nordiska Ministerrådet, 1991).

The energy taxes and the carbon dioxide tax generate important revenues, while the sulfur tax contributes small amounts from a government budgetary perspective. In 1992, revenues from the carbon dioxide tax were around SEK 9 billion, while the sulfur tax contributed just below SEK 300 million (Ministry of Environment and Natural Resources, 1993). The nitrogen oxides charge does not generate any general revenue at all, as will be explained shortly. The differentiation of the energy tax on diesel fuels was not meant to affect tax revenue in a major way, but due to the unexpected rise in sales of the environmentally better quality fuels, tax revenues have in fact dropped. If market shares for fuel classes 1, 2, and 3 remain at present levels (15%, 60%, and 25%, respectively), it is estimated that the loss of government revenue will be around SEK 600 million per year compared to

Table 5.1. Economic instruments for air pollution control in Sweden, October 1993.

	Year of introduction	Level and scope	
Carbon dioxide tax	1991	Industry Other sectors	SEK 80/tonne CO_2 SEK 320/tonne CO_2
Sulfur tax	1991	All sectors	SEK 30,000/tonne S
Nitrogen oxides charge	1992	Large and medium-sized boilers	SEK 40,000/tonne NO_x
Differentiated sales tax on light vehicles	1993	Reduction compared to class 3: Env.class 1 Env.class 2 Env.class 3	 SEK 6,000/vehicle SEK 2,000/vehicle SEK 0
Differentiated tax on diesel fuels	1991	Reduction compared to class 3: Env.class 1 Env.class 2 Env.class 3	 SEK 535/m^3 SEK 250/m^3 SEK 0
Differentiated petrol tax	1986	Tax reduction for lead-free petrol	SEK 0.51/liter

1 SEK = 0.11 ECU = 0.13 US$.

a situation where only class 3 (the standard grade) were provided (Swedish Environmental Protection Agency, 1993c).

Tests on motor vehicles have revealed that lower emission levels than those currently required can be achieved. In response to the need for further emission reductions, the Swedish parliament has passed a tax incentive scheme with voluntary emission targets allowing a choice between three levels of standards (environmental classes). This will permit the gradual introduction of low-emission vehicles in Sweden. Tax incentives for light vehicles became effective in 1993, and similar incentives for heavy vehicles are scheduled for 1994. As in many other countries, lead-free petrol is taxed at a lower rate than leaded petrol. The Swedish Environmental Protection Agency has recently proposed a broader tax incentive scheme designed not only to encourage the phasing out of leaded petrol, but also to improve the properties of petrol in general.

This chapter elaborates on those schemes that have been effective long enough for responses to be recorded. Thus, the recent environmental classification of vehicles is excluded, as well as an analysis of the carbon dioxide tax, since this would have to deal with energy taxes as well. Before looking at each economic instrument, a few words on the Swedish energy structure, the environmental situation, and the development of emissions may be helpful.

5.2 Swedish Energy Structure

Energy-intensive processes based on natural resources such as forests, iron ore and hydropower are important within Swedish industry. Following the oil crises of the 1970s energy efficiency improved substantially, and the share of oil in total energy supply dropped from 70% to 45% between 1970 and 1990. The use of nuclear power has increased (Swedish Board for Industry and Technical Development, 1991), but the Swedish parliament has decided that the existing nuclear reactors are to be phased out, a process that will be completed by the year 2010.

At present, Swedish electricity production is based almost exclusively on hydro and nuclear power, each source accounting for approximately half the electricity supply. Due to the severity of the Swedish climate the demand for electricity for domestic heating is high. In high-density residential areas a large part of this demand is covered by district heating.

5.3 Environmental Problems and Emissions

Deposition of sulfur exceeds critical loads in most parts of Sweden. In the south of the country the deposition of sulfur needs to be reduced by 60%–80%. Nitrogen deposition also exceeds critical loads, especially in the south, where deposition should be reduced by around 60% (*Table 5.2*). Foreign sources dominate both sulfur and nitrogen deposition in Sweden; the Swedish contribution to the deposition of sulfur is only around 10% and that of nitrogen around 20%.

Emissions of sulfur dioxide in Sweden have been decreasing since the early 1970s, and are now lower than they were in the 1950s. Total emissions have decreased by almost 80% since 1980. In contrast, total Swedish emissions of nitrogen oxides decreased very little between 1980 and 1991. The emissions from stationary sources have been reduced by almost 50%,

Table 5.2. Swedish contribution to sulfur and nitrogen deposition and reduction required.

	Reduction necessary to avoid exceeding critical loads (%)	Swedish contribution to deposition (%)
Sulfur	60–80	10
Nitrogen	60	20

Source: Swedish Environmental Protection Agency, 1994.

but this reduction has been largely offset by the growth of emissions from the transport sector (Swedish Environmental Protection Agency, 1994).

5.4 The Sulfur Tax

The sulfur tax, which came into force in 1991, is levied on coal, peat, and oil consumption and corresponds to SEK 30,000 (approximately US$ 3,900) per tonne of sulfur emitted. The tax aims at a cost-effective reduction of Swedish sulfur emissions by 80% before the year 2000. The sulfur tax has not replaced administrative regulations, but is complementary to them. As a result, the tax coexists with emission standards and individual permits for large stationary sources. When the tax was designed, it was anticipated that administrative regulations would also have to be tightened to meet the national target for the late 1990s. The tax was intended to achieve reductions of approximately the same magnitude as those expected from stricter standards, but to achieve them more rapidly.

Swedish sulfur emission standards apply to all fuels, and are being tightened in line with the original plans. The general emission limits will be 0.05 g S/MJ fuel input (annual average) for combustion plants emitting more than 400 tonnes of sulfur per year, and 0.10 g S/MJ for smaller plants (annual average). The short-term maximum emission limits will remain at 0.19 g S/MJ, corresponding to 0.8% sulfur in heavy fuel oil. The new standards came into force in large parts of southern Sweden in 1993 and will apply to all parts of the country by 1997. In addition, the limit for the sulfur content in gas oil was reduced from a maximum of 0.3% to a maximum of 0.2% as of 1993 (annual average limit).

5.4.1 Design of the tax system

The sulfur tax is designed as a fuel tax; the tax rate levied depends on the sulfur content of the fuel. For practical reasons, oils containing less than

0.1% sulfur are not taxed. The sulfur tax is refunded to the extent that emissions have been controlled by flue gas desulfurization or other cleaning technologies. Thus, the tax encourages both the use of low-sulfur fuels and investments in sulfur emission control. Fuels used by shipping or for purposes other than energy generation (such as raw materials in industrial processes) are not taxed. The tax revenue is part of general government revenue.

The rather high rate of the tax must be viewed in relation to earlier sulfur control policies and the ambitious emission reduction target. When the tax was introduced the sulfur content of fuel oil was already low, as were average sulfur emissions from coal-fired boilers. The sulfur tax was designed to bring about further decreases from these low levels.

5.4.2 Results

Total Swedish emissions of sulfur dioxide from combustion in stationary sources decreased from 54,000 tonnes in 1990 to 40,000 tonnes in 1991, a decrease of about 25% (Statistics Sweden, 1993). Several factors contributed to this, including the sulfur tax and the administrative regulations concerning sulfur, the carbon dioxide tax and other energy taxes, and changes in the level of industrial activity. I will not go into the changes in the overall pattern of energy use related to changes in energy taxes and industrial activity, but concentrate on distinguishing the effects of the sulfur tax and the administrative regulations on more direct measures to reduce sulfur emissions, such as the choice of fuel qualities and various emission reduction measures.

In 1990, large users of *heavy fuel oil* in some cities had already begun to use very low sulfur fuel, due to local restrictions or on a voluntary basis. Consequently, the actual average content of sulfur in heavy fuel oil was around 0.65%, somewhat lower than the maximum value. The introduction of the sulfur tax and the tightening of the general emission standards for combustion plants has brought the average sulfur content of heavy fuel oil down to around 0.4%.

The effects of the introduction of the tax (in 1991) and the subsequent tightening of emission standards (starting in 1993) cannot be clearly isolated in this case. The combined effect has been a very rapid switch to higher-quality heavy fuel oil containing approximately 0.4% sulfur in all parts of the country, not just in those areas where the tighter emission standards have come into force. Low sulfur fuels gained large market shares as early as 1991. In view of this, it seems likely that much the same changes could have been brought about by the sulfur tax without the subsequent tightening of emission standards. However, the very quick convergence to the new standard quality was probably helped by the combination of instruments.

In the case of *gas oil*, the specific limit for the sulfur content (maximum 0.2%) is the most stringent administrative requirement. Gas oil now contains less than 0.1% sulfur, so the effects of the sulfur tax have clearly been dominant. The sulfur content of diesel fuels has been reduced even further, although this has been mainly due to another incentive mechanism – the differentiation of the tax on diesel fuel. As has already been pointed out, the sulfur tax does not apply to oils whose sulfur content is below 0.1%.

The use of *coal and peat* is rather small in the Swedish energy sector. Coal- and peat-fired installations are faced with strict emission control requirements in their permits under the Environmental Protection Act. These permit requirements were very stringent even before 1993, so the tightening of the general emission standards hardly affected the coal- and peat-fired installations. New emission reduction measures have been taken, however, to improve the efficiency of sulfur removal at most installations and to increase the use of low-sulfur coals. These emission reduction measures can in most cases be ascribed to the sulfur tax. The introduction of stricter requirements in individual permits has only been important for some coal-, peat- and oil-fired installations, since only a small fraction of permits is reviewed each year. For the vast majority of combustion plants, the requirements for sulfur emissions in individual permits has not changed in the last few years.

In summary, the sulfur tax seems to have been decisive in reducing emissions from the combustion of coal, peat and gas oil in stationary sources. As for heavy fuel oil, the rapid convergence to a new standard quality was probably achieved by the combination of economic and administrative instruments.

When the sulfur tax rate was set, it was based upon assessments of the marginal cost of reducing sulfur emissions in Sweden. The cost of reducing sulfur emissions from the combustion of oil is very important in this context, since it is dependent, among other things, on the world market prices of very low sulfur oils in relation to other oils. The tax rate was set so as to provide some room for increasing the relative prices of very low sulfur oils. So far, however, relative prices have not changed to a great extent. Low sulfur fuels have been available at fairly low cost. Therefore the average cost of the emission reduction measures undertaken so far is probably as low as SEK 10,000 per tonne of sulfur abated.

The administration of the sulfur tax has proved fairly simple. The same administrative framework is used to handle the taxes on energy, sulfur, and carbon dioxide. The sulfur content of fuels had to be measured even before 1991 to ensure compliance with existing regulations. Approximately 65 plants per year have been reimbursed due to their sulfur reduction measures. In 1992, eligibility for reimbursement was based on continuous monitoring of emissions in 80%–90% of cases. Since the requirements for monitoring both SO_2 and NO_x emissions are similar (see Section 5.5), and plants with sulfur reduction measures are liable to the nitrogen oxides charge, it has been possible to coordinate monitoring of sulfur and nitrogen oxides emissions and limit administrative costs.

5.5 The Nitrogen Oxides Charge

The nitrogen oxides charge aims at the cost-effective reduction of Swedish nitrogen oxides emissions from combustion plants. Since vehicles are the main source of nitrogen oxides emissions, the nitrogen oxides charge affects a rather small portion of Swedish emissions. The importance of this charge will increase when new electricity power plants are built, possibly by the end of the 1990s. The national target implies that total Swedish nitrogen oxides emissions will need to be reduced by 30% by 1995.

Many of the plants affected by the nitrogen oxides charge must also meet requirements laid down in individual permits, but there are no general emission standards of the type applied to sulfur. A tightening of the requirements of the permits was foreseen when the new economic instrument was introduced. The nitrogen oxides charge was set so as to achieve roughly the same decrease as that anticipated from tightened permit requirements, but more rapidly. Since then, permits have been reviewed in a number of cases but, as was explained in relation to sulfur, changes in permits affect a fairly small fraction of installations in any one year.

5.5.1 Overall design of the charge system

The nitrogen oxides charge came into force in 1992, based on a decision made as early as 1990. The charge is levied on large and medium-sized combustion plants (boilers with a capacity of over 10 MW, producing more than 50 GWh per year). In 1992 approximately 120 plants (representing 185 boilers) were affected by the charge. The charge is SEK 40,000 (about US$ 5,200) per tonne of NO_x emitted.

To avoid giving smaller plants an advantage, the revenue from the nitrogen oxides charge is refunded to the group of plants that pay it, in proportion to the amount of useful energy each plant produces. Thus, the system creates incentives both to reduce emissions and to increase energy efficiency. Plants with comparatively low emissions will be able to make a profit, while those with comparatively high emissions will have to make net payments. This does not harm the incentive effects of the charge, as the amount repaid to each plant is related to the amount of useful energy produced and is thus independent of its efforts to reduce emissions.

5.5.2 Monitoring

The nitrogen oxides charge was the first charge to be based on measured emissions. Emissions of nitrogen oxides can either be measured continuously or calculated from a fixed level of 250 mg/MJ for boilers and 600 mg/MJ for gas turbines. Since actual emissions are much lower in most cases, due to environmental regulations, continuous monitoring is generally preferable. The monitoring system must meet requirements on equipment, data recording and data processing specified by the Swedish Environmental Protection Agency. During 5% of the monthly operating time, emissions may be estimated on the basis of emission values achieved under similar operating conditions. This allowance is made to ensure proper time to calibrate and maintain the system. If the system is out of order or does not comply with the requirements, emissions are estimated on the basis of the fixed emission levels of 250 mg/MJ for boilers and to 600 mg/MJ for gas turbines.

Monitoring equipment is a large item of expenditure. The annual cost of monitoring nitrogen oxides emissions is estimated at SEK 300,000 per plant, which is why the charge is not levied on very small plants. For most plants, monitoring would have been required to ensure compliance with existing environmental protection legislation even if the charge had not been introduced.

5.5.3 Results

The nitrogen oxides charge has already had considerable impacts on emissions of nitrogen oxides from the plants which are part of the system. The methods applied for reducing emissions can be divided into three groups: (1) combustion improvement measures; (2) selective noncatalytic reduction (SNCR, involving the injection of urea or ammonia); and (3) flue gas cleaning by selective catalytic reduction (SCR).

Combustion measures have been implemented at almost every facility. In general, these are the cheapest and most cost-effective measures, including low-NO_x burners, flue gas recirculation, air staging, reburning, and "fine tuning" of the combustion system. In many cases, simply running the plant in an optimal way will cut emissions. To achieve this, some plants offer bonus systems to their employees, where the size of the bonus increases as emissions decline.

No Swedish plants had installed SNCR (injection of urea or ammonia) before the decision was made to introduce the NO_x charge. This has proved to be an inexpensive method of reducing emissions. Today, about 20 plants have installed SNCR, and the method is being considered for several other plants (Energigruppen Syd, 1993; Hjalmarsson, 1992). Most of the units affected by the charge are small by international standards, so that SCR has been used in only a few cases. SCR has been the most cost-effective way to reduce emissions for some coal-fired boilers (over 200 MW_{th}) with high initial emissions.

Total emissions from the plants liable to the charge fell from about 25,000 tonnes in the late 1980s to about 15,000 tonnes in 1992, a reduction of roughly 40% (Swedish Environmental Protection Agency, 1993b). Permits have been tightened in a few cases, but for most plants the requirements have not been changed in the last few years. Roughly 50% of the plants have no requirements at all as to NO_x reduction in permits (Energigruppen Syd, 1993). Thus, the nitrogen oxides charge must be considered to have been the major driving force behind the emission reductions.

The average cost of combustion measures taken so far is estimated at SEK 10,000–15,000 (US$ 1,300–2,000) per tonne NO_x abated for energy production, waste incinerators, and combustion plants in the pulp and paper industry, while the corresponding cost for combustion plants in the chemical industry is estimated at SEK 20,000 per tonne.

The average cost of cleaning measures is higher, estimated at SEK 15,000–25,000 NO_x abated for waste incinerators and combustion plants in the pulp and paper industry. For energy production the estimated average cost is SEK 35,000 per tonne. The latter cost is dominated by the installation of "high-dust" SCR for three coal-fired boilers (Energigruppen Syd, 1993). All the cost figures quoted for NO_x abatement are based upon 10 year lifetimes of investments and an interest rate of 11%.

The cost of continuous monitoring of NO_x emissions is estimated at SEK 4,000 per tonne NO_x abated. However, many plants would have incurred monitoring costs even in the absence of the charge.

Practically all waste incinerators have had to make net payments as a result of the charge system. Within industry some plants lose and others gain. The same applies to the production of electricity and heat (Swedish

Table 5.3. Emissions, energy production, and payments related to the nitrogen oxides charge in 1992.

	NO_x emissions (tonnes)	Useful energy produced (GWh)	Payment of charge (SEK million)	Repayment (SEK million)	Net result (SEK million)	No. of plants
Waste incineration	1,097	1,690	43.9	27.3	–16.5	5
Energy production	8,266	21,628	330.6	349.6	19.0	53
Chemical industry	1,278	3,184	51.1	51.4	0.3	23
Metal industry	77	469	3.1	7.4	4.3	2
Pulp & paper ind.	4,587	10,503	183.5	169.8	–13.7	39
Total	15,305	37,465	612.2	605.6[a]	–6.6	122

[a]SEK 5 million was not paid back at the initial repayment but was kept as a reserve pending final revision of the accounts. SEK 1.6 million was used to cover administrative costs.

Source: Swedish Environmental Protection Agency, 1993b.

Environmental Protection Agency, 1993a, 1993b). On average, waste incinerators and the pulp and paper industry have made net payments, while energy producers and, to some extent, the metal industry have received refunds (*Table 5.3*).

The administration of the nitrogen oxides charge is more complex than that of the sulfur tax due to the repayment of charge revenues. Still, the administration costs to the Environmental Protection Agency were less than SEK 200 per tonne NO_x abated in 1992. Most boiler owners have a positive attitude to the system; it is considered fair because it gives bonuses to plants with low emissions.

A government commission has recently proposed that the charge system should be extended to cover even smaller combustion plants and some industrial processes (Anonymous, 1993). According to these proposals, the charge threshold would be reduced from a minimum annual energy production of 50 to 25 GWh, irrespective of the size of the boiler. This would add another 500 boilers to the charge system. In addition, some industrial processes would be subject to the charge. A parliamentary decision on these proposals is expected during 1994.

5.6 Differentiation of the Tax on Diesel Fuels

In January 1991 the energy tax on diesel fuels was differentiated to encourage the use of environmentally better grades. Diesel fuels with "standard" specifications belong to environmental class 3, those with better specifications to class 2, and the best grades to class 1. The environmentally better grades are taxed at a lower rate than the standard grade.

The purpose of the differentiation of the tax on diesel fuels is to improve air quality in urban areas, where vehicle emissions are the predominant source of the health risks. The catalytic converters now used in petrol-powered vehicles probably greatly reduce emissions of carcinogenic substances from these vehicles. In future, attention will therefore have to focus on reducing emissions of carcinogenic particles and gases from diesel-powered vehicles. The concentrations of sulfur dioxide in Swedish urban areas are lower that in many other countries. Historical monuments are still being attacked by air pollution particularly in the form of sulfur dioxide, but concentrations of sulfur dioxide no longer present major health risks.

5.6.1 Design of the system

Improvements in the quality of diesel fuel contribute to the reduction of health risks and corrosion in urban areas. Problems of smoke and odor are also lessened. The environmental classification of diesel fuels is based on a number of parameters (see *Table 5.4*). A low sulfur content gives rise to lower emissions of sulfur and particles and is essential for the efficiency of particle filters and catalytic converters in cleaning exhaust gases. Low contents of aromatic hydrocarbons, including polycyclic aromatic hydrocarbons (PAH), mean reduced emissions of substances that are hazardous to human health. The distillation range, the cetane number (a measure of the ignitability of diesel fuel), and density are significant factors with regard to both emissions and engine running characteristics.

The tax differentials have been set so as to offset the extra cost of the environmentally better grades of fuel. The system was introduced in 1991. Definitions of class 1 and class 2 fuels were upgraded in January 1992, and the tax differentials were adjusted. In October 1993, the taxation of diesel-powered vehicles was reformed, to replace the so-called kilometer tax by a new diesel tax. Simultaneously, the tax reduction for class 1 fuels was increased. *Table 5.4* shows the tax differentials in January 1992 and October 1993. Apart from the energy tax and diesel tax, diesel users also pay a carbon dioxide tax (see *Table 5.5*), which is the same for all fuel classes. The diesel tax is also the same for all fuel classes.

Table 5.4. Environmental classification of diesel fuels and tax differentials.

	Class 1	Class 2	Class 3
Sulfur (mass content) (max, %)	0.001	0.005	0.2[a]
Aromatics (volume content) (max, %)	5	20	
PAH (volume content) (max, %)	0	0.1	
Cetane number (min)	50	47	
Density (kg/m^3)	800–820	800–820	
Distillation:			
Initial boiling point (min, °C)	180	180	
Temp. at 95% recovery (max, °C)	285	295	
Tax reduction per liter compared to class 3:			
January 1992			
SEK	0.45	0.25	
ECU	0.048	0.027	
October 1993			
SEK	0.535	0.25	
ECU	0.057	0.027	

[a]Limit value; the actual sulfur content is generally below 0.1%.

Table 5.5. Taxes on diesel fuels used by road vehicles in Sweden as of October 1993 (per m^3).

		Class 1	Class 2	Class 3
Energy tax	SEK	5	290	540
	ECU	0.5	31	57
Carbon dioxide tax	SEK	920	920	920
	ECU	98	98	98
Diesel fuel tax	SEK	1,300	1,300	1,300
	ECU	138	138	138
Total taxes	SEK	2,225	2,510	2,760
(excluding VAT)	ECU	237	267	294

5.6.2 Results

The differentiation of the tax on diesel fuels has affected markets very strongly (Swedish Environmental Protection Agency, 1993c). In 1990, less than 1% of diesel oil would have met the current requirements of class 1 and class 2 fuels. In the first part of 1993, 75% of diesel fuels sold were classified as class 1 or 2 (15% as 1 and 60% class 2). Initially, most of the environmentally better fuels were imported but now some domestic refineries have invested in new facilities to produce their own fuels of classes 1 and 2. The tax differentiation has encouraged research and development and Sweden is now at the forefront in the production of environmentally better diesel fuels.

Sulfur emissions from diesel vehicles have dropped by approximately 75%. Emissions of carcinogenic substances have also been reduced, but at the present state of knowledge this reduction cannot be quantified. There has also been some reduction in nitrogen oxides emissions, which will contribute to lower incidences of cancer, allergies, and corrosion in urban areas.

The cost to society in the form of higher production costs in domestic refineries or more expensive oil imports has been estimated at SEK 430 million per year. The extra costs of the environmentally better grades have so far been lower than expected when the tax differentials were decided. One reason is that class 1 and class 2 diesel fuels require production processes similar to those for aviation fuel. The demand for aviation fuel has fallen in the last few years, so that the refineries have been able to use their spare capacity to produce class 1 and class 2 diesel fuels (Swedish Environmental Protection Agency, 1993c). The administration of the system has worked without any major problems. The risk of misuse of the system is regarded as limited as the oil companies fear bad publicity. The oil companies are positive to the system as it gives them freedom to choose the qualities of fuel they supply.

5.7 Conclusions

Economic instruments have achieved a breakthrough in Sweden. The sulfur tax, the nitrogen oxides charge, and the differentiated tax on diesel fuels have resulted in sharp reductions in emissions. Many low-cost measures have been found and the new economic instruments appear to be encouraging innovation and technical change. For example, combustion measures and SNCR techniques have been applied in new ways to reduce emissions of nitrogen oxides. In short, the new instruments have so far lived up to their theoretical potential.

In general, the economic incentive mechanisms have been well accepted. The full repayment of the nitrogen oxides charge as well as the technique of differentiating an existing tax, have contributed to this without increasing the total tax burden. The Swedish government has affirmed the important role of market mechanisms in environmental policy, as can be concluded from several government bills recently approved by parliament (Ministry of Finance, 1992; Ministry of Environment and Natural Resources, 1993). New economic instruments will no doubt be introduced as complements to the traditional regulations.

Acknowledgments

Many individuals at the Swedish Environmental Protection Agency have contributed to this paper. I am grateful for their help, as well as for helpful comments from the referees.

References

Anonymous, 1993, *Incentives and Controls for a Better Environment: Proposals Concerning More Extensive Use of Economic Instruments in Controlling Emissions of Nitrogen Oxides*, SOU 1993:118, Stockholm, Sweden (in Swedish).

Energigruppen Syd, Energikonsult A. Bauer AB, 1993, *Assessment of the Nitrogen Oxides Charge*, Malmö, Sweden (in Swedish).

Hjalmarsson, A.K., 1992, *Scandinavian Plants with NO_x Control Installations*, Swedish Environmental Protection Agency Report 4152, Solna, Sweden.

Ministry of the Environment and Natural Resources, 1993, *The Environment: Facts and Future*, Offprint from Government Bill 1992/93:100, Annex 100, Stockholm, Sweden.

Ministry of Finance, 1992, *Economic Instruments in Environment Policy*, Offprint from Government Bill 1991/92:100, Annex 15, Stockholm, Sweden.

Nordiska Ministerrådet, 1991, *Energy and Environment in the Nordic Countries: Taxes and Charges as Environmental Instruments*, Nord 1991:23, Copenhagen, Denmark (in Swedish).

Statistics Sweden, 1993. *Sulfur Dioxide, Nitrogen Oxides and Carbon Dioxide Emissions to Air in Sweden in 1991*, Stockholm, Sweden.

Swedish Board for Industrial and Technical Development, 1991, *Energy in Sweden: Facts and Figures*, Stockholm, Sweden.

Swedish Environmental Protection Agency, 1993a, *Environmental Charges on the Emissions of Nitrogen Oxides from Combustion Plants*, 1993-05-03, Solna, Sweden (in Swedish).

Swedish Environmental Protection Agency, 1993b, *Information Concerning the Nitrogen Oxides Charge*, 1993-08-27, Solna, Sweden (in Swedish).

Swedish Environmental Protection Agency, 1993c, *Environmentally Differentiated Taxes on Diesel Oils: The Swedish Experience*, Solna, Sweden.

Swedish Environmental Protection Agency, 1994, *Strategy for Sustainable Development: Proposals for a Swedish Programme*, Solna, Sweden.

Chapter 6

Air Pollution Control in the United States: A Mixed Portfolio Approach

Nancy Kete
US Mission to the OECD
Paris, France

Abstract

Much has been made of the experiments with emissions trading and other forms of incentive-based regulation in the United States. This chapter takes a step back to outline the principal program elements of the Clean Air Act, and then looks retrospectively at some of the issues that have arisen during the past two decades of debate on the role of emissions trading. It then goes on to differentiate that experience from new incentive-based programs created or authorized by the 1990 amendments, especially the acid rain control program. In considering the qualities of the acid rain program, or any other regulatory regime, the administrative structure cannot be separated from the primary objectives. The qualities that support a functional emissions trading regime do not inhere in using incentive-based regulation. They result from self-conscious, transparent decisions to build such qualities into whatever program is chosen.

Key words: acid rain, emissions trading, Clean Air Act, regulation, economic incentives, pollution control.

6.1 Introduction

Since at least the late 1960s, the United States has supported a cottage industry of experts on the advantages of economic instruments for pollution control. From the theoretical literature to public policy debates, anyone involved or interested in environmental protection has been exposed to the idea of cost-effective regulation through emissions trading and related policy instruments. Yet, while the case for what I prefer to call incentive-based regulation has been compelling in the abstract, in practice political and practical realities have resulted in a policy "system" that includes a range of public interventions in the economic system's air pollution behavior. Although for years the debate seems to have been polarized around two alternative concepts – command-and-control on the one hand, and economic instruments on the other – in reality policy incorporates options that lie along a continuum of responses to address air pollution problems. The air pollution control system under the Clean Air Act (CAA) has multiple objectives and multiple response strategies to achieve them. One thesis of this chapter is that there is no such thing as an "optimal" air pollution policy in the sense the economists and systems analysts usually mean. At best policy makers strive for noninferior sets – policy solutions that are not obviously worse than alternative feasible options. Policy makers and regulators have used and will continue to use a mix of policies and instruments to reduce air pollution problems, and to do so in politically and economically acceptable ways. The exact mix may change over time, and what will be considered feasible or acceptable will evolve, but the mixed portfolio of responses is here to stay (Hawkens, 1981). The continuum of responses includes, for example information-based programs; government–industry collaboration; subsidy removal and regulatory reform; environmental taxes; marketable permits; emissions trading; emission limits; and best-technology requirements. Each instrument is equally legitimate although not equally appropriate under every circumstance (Kete, 1994).

Emissions trading and economic instruments advocacy has been weakened by its often doctrinaire recommendations and criticism of the status quo. For all the attention given to the theoretical advantages of alternative policy instruments, far too little has been paid to the institutional and administrative details that must be worked out to make any program – market-based or otherwise – work. Worse, what began in the 1970s as legitimate regulatory reform ideas became during the 1980s regulatory relief. Potentially constructive recommendations were thus politicized and tainted. But the past few years have seen a renewed interest in regulatory reform among environmentalists and regulators who were not keen on economic instruments before the 1990 Clean Air Act amendments, and others. What changed and why? First, regulators have learned lessons from their experiences to date with economic instruments, including the work to develop the acid rain program legislation, the implementation of the lead trading program, and the phasing out of CFCs under the Montreal Protocol. Regulators and their critics also learned important lessons about the limits of incentive-based approaches through, for example, the unfortunate experiences associated with the "bad" bubbles of the past, under the EPA's Emissions Trading Policy Statement. Second, the new CAA is, we hope, better constructed to support an increase in the use of economic instruments to help regulators meet the Act's goals and to lower the cost of doing so.

The question of whether a statutory goal could be achieved (or achieved more effectively) under a regime that includes economic instruments is an empirical one. Unfortunately, there is little actual experience with either of the two distinct categories of incentive-based regulation – emission taxes and marketable permits – and only a mixed record with a set of mechanisms known collectively as "emissions trading", which includes bubbles, netting, banking and offsets. Even the acid rain control program with its promising emissions allowance system remains an incomplete experiment in regulatory reform. It remains to be seen whether market-based regulation can achieve public ends better than the alternatives.

This chapter discusses the institutional aspects of air pollution control, and looks at emissions trading in this context. Economic instruments, like all regulatory instruments, are means to an end, or ends. Thus, rather than begin with a discussion of the use of economic instruments under the Clean Air Act, this chapter takes the Act itself as the first subject of its inquiry. Section 6.2 briefly describes the general framework and motivation for the US Clean Air Act in terms of the legal and social relations embedded within it. Section 6.3 describes the environmental quality objectives (EQOs) at the heart of the Act. Section 6.4 takes the reader from the environmental quality objectives of the Act to the practical way in which these objectives are to be achieved. Section 6.5 discusses the persistence of the problem of failure to meet the EQOs in many areas of the USA. Section 6.6 describes the best available technology requirements and how they relate to the EQO described in Section 6.3. Section 6.7 begins a discussion of emissions trading, describing the development and controversies over the Emissions Trading Policy Statement. Section 6.8 looks at the controversy over those aspects of emissions trading related to new source control requirements. Section 6.9 considers the acid rain program as regulatory reform and describes the new acid rain control program and its emission allowance trading system. Section 6.10 provides a discussion and summary.

6.2 The Clean Air Act[1]

The Clean Air Act (CAA) is a normative response to a problem of social cost. In brief, the purpose of the CAA is to protect and enhance the quality of the nation's air so as to promote the public health and welfare and the productive capacity of its population. This overall objective is (or is meant to be) accomplished through a complex web of interrelated requirements, including environmental quality objectives; state-developed plans to ensure the environmental quality objectives are met; source-specific emission limits established under the state plans; several types of technology-based standards, and finally, for electric utilities, the emission allowance system for acid rain control.

The CAA and its implementing regulations comprise an institution – a set of working rules – replete with mutualities and dualities, rights and duties, privileges and exposures. Essentially, the CAA entitles the public to breathe clean air and to enjoy an environment that has not been degraded by air pollution; polluters thus have a corresponding duty to honor that entitlement. However, most of the Act's requirements are mediated by the EPA administrator's (and ultimately, the federal courts') interpretation of what they mean. Nonetheless, it is clear that the social relation embedded in the CAA is one of public control over the assimilative capacity of the air. This is a definite, albeit gradual, change in the rights and duties respecting environmental quality and assimilation of environmental services. Historically, industry, publicly owned sources of air pollution, and individuals as owners and operators of automobiles and wood stoves, for example, have claimed privilege over the dilution services provided by the ocean of air in which we live, leaving the general public without power to prevent any adverse consequences. Since 1970, the Clean Air Act and subsequent amendments have steadily shifted the relationship between polluter and others, providing more, and more effective, control over these activities to ensure the achievement of the right to clean air and a healthy environment. Still, for the most part, emitting sources generally have no direct legal duty

[1]The discussion that follows is offered as a nontechnical, nonlegal guide to the workings of the major elements of the Clean Air Act. The reader interested in the text of the law itself, and a comprehensive presentation of its history, structure and interpretation of each major section should see the *Clean Air Deskbook* (1992).

to undertake pollution control or pollution prevention efforts beyond the Clean Air Act's indirect requirements. (As discussed below, the acid rain program's emission reduction requirements are a notable exception to the general case.) Thus despite the nominal entitlement to a clean and healthy environment, effectively the CAA imposes on the private sector a general duty to behave in accordance with regulations promulgated by the administrator. These regulations are (at least theoretically) the institutionalization of the general entitlement to clean air and a safe environment provided for by the law.

Once the administrator has promulgated regulations, the implementation of many of the Clean Air Act's requirements shifts to the states under an essentially federalist construction. Furthermore, most of the Clean Air Act's provisions establish federal minima with the right to go farther toward protecting public health and welfare reserved to the states and localities. Section 6.116 of the Clean Air Act stipulates that nothing shall preclude or deny the right of any state or political subdivision thereof to adopt or enforce (1) any standard or limitation with respect to air pollution, or (2) any requirement with respect to control or abatement of air pollution. As a practical matter, however, there has not been a lot of experience with states going beyond the federal minima.

6.3 Environmental Quality Objectives

How, specifically, are the rights and the duties of the Clean Air Act operationalized? At the heart of the Clean Air Act are its environmental quality objectives, the national ambient air quality standards (NAAQS). The NAAQS essentially are critical levels of concentrations of air pollution below which no adverse effects are presumed to occur. The NAAQS-setting process is science-based and generally takes upwards of ten years. At present there are six pollutants for which NAAQS have been established: sulfur dioxide, nitrogen oxides, particulate matter, lead, ozone, and carbon monoxide. The NAAQS are set in terms of concentration, e.g., parts per million (ppm) or micrograms per cubic meter ($\mu g/m^3$) during a specified measurement period, e.g., annual arithmetic mean or maximum 24-hour concentration. The primary NAAQS are set to protect public health, with an adequate margin of safety and the secondary NAAQS are set to protect public welfare. The NAAQS apply everywhere in the USA, that is, they are invariant with geography.

6.4 From EQOs to Emissions Control: The State Implementation Plan

At this point the states come in. The CAA requires each state to formulate a State Implementation Plan (SIP) that is adequate to demonstrate the attainment and maintenance of each NAAQS. The principal function of the SIP is thus to ensure, through air quality modeling and monitoring, and ultimately the setting of enforceable emission limits, that sources of emissions individually and collectively do not cause pollution, defined as an exceedance of the NAAQS. The Act requires that the SIP provide for the implementation, maintenance, and enforcement of the NAAQS. At a minimum, the Act requires that a SIP include: enforceable emission limitations and other control measures for sources; schedules and timetables for compliance; appropriate measures for monitoring and modeling ambient air quality and emissions from stationary sources; and other specifics. The enforceable emission limitations, or "SIP limits", effectively specify or operationalize each source's duty to protect the public's right to be free of polluted air. Once approved by the EPA administrator, the SIPs are enforceable under both federal and state law. The NAAQS/SIP system can be considered, at least in theory, the first-line defense against air pollution. Yet we see already how indirect is the relationship between the primary goal of the Clean Air Act and any operative constraints on a source's behavior.

6.5 The Persistent Problem of Nonattainment of the NAAQS

A notable characteristic of the system which I have just described as the heart of the Clean Air Act, that is, the combination of the federal NAAQS with the State Implementation Plans, is its failure to meet certain key objectives. Most major American cities and metropolitan areas experience one or more exceedances of one or more of the NAAQS each year with roughly half of the US population estimated to be exposed to harmful levels of air pollution. Areas that fail to meet one or more NAAQS are called "nonattainment areas". (An area may be in nonattainment for one pollutant, such as ozone (photochemical smog), and in attainment for all the other five NAAQS.) There are numerous reasons for this failure, not least of which was the overoptimism on the part of Congress in 1970 and again in 1977 that attainment and maintenance of good air quality would be achievable given the tools it made available under the Clean Air Act. Second, despite an aggressive mobile source emission control program, the continual increase in the number of automobiles, the number of miles they travel, and the amount of traffic congestion all contributed to an increase in the amount of pollution coming from mobile sources. Despite dramatic decreases (60%–80% less compared to cars of the 1960s) in the amount of pollution emitted from each vehicle, congestion and growth in vehicle use have overwhelmed the benefits of cleaner cars. In addition, the law's exceptions for small, diffuse "area" sources put potentially beneficial controls out of the reach of regulators. In 1977, the Congress amended the Clean Air Act to create, among other things, a special section establishing requirements for nonattainment areas (part D), but still, despite success in reducing lead and sulfur dioxide concentrations, ozone, CO and PM-10 nonattainment problems persist.

6.6 Best Available Technology Requirements

In addition to the NAAQS and related SIP provisions, the CAA imposes several types of technology-based standards on new and modified sources of emissions. The new source performance standards (NSPS) provisions under section 111 of the Act require the EPA to establish national emission standards by source category (for example, electric utility power plants) for each pollutant that causes or contributes significantly to air pollution which may reasonably be anticipated to endanger public health or welfare. The legislative history of the 1970 amendments makes clear that the purpose of this section "is to prevent the occurrence anywhere in the United States of significant new air pollution problems arising from such sources either because they generate extra hazardous pollutants or because they are large scale polluters" (CRS, 1974). The legislative history further shows that the NSPS are meant to harmonize requirements to avoid environmentally destructive competition across the several states: "The promulgation of federal emissions standards for new sources ... will preclude efforts on the part of states to compete with each other in trying to attract new plants and facilities without assuring adequate control of extra hazardous or large-scale emissions therefrom" (CRS, 1974). The Senate Committee Report added, in an argument in favor of long-term management of air quality resources: "standards of performance should provide an incentive for industries to work toward constant improvement in techniques for preventing and controlling emissions from stationary sources, since more effective emission control will provide greater latitude in the selection of sites for new facilities" (CRS, 1974, Vol. 1, p. 417). For these reasons, an additional relationship embedded in the CAA is the requirement that new sources be clean. Again, however, the definition of "clean" must be operationalized by the EPA in setting national standards.

A second type of technology-based standard applicable to new sources applies on a case-by-case basis under the new source review program. The new source review program has two parts: nonattainment area New Source Review (NSR) and the Prevention of Significant Deterioration (PSD) program under part C of the Act. Under both PSD and NSR the starting place for making the case-by-case determination of what is the best technology that should be applied to the facility in question is the NSPS for the source category. For a source proposed to be located in an attainment area where the NAAQS are met, the PSD provisions apply and the permitting process must determine (among other things) what the best available control technology (BACT) should be. For a source proposed to be located in a nonattainment area the nonattainment new source review requirements apply, and, among other things, the source is required to meet LAER – the lowest achievable emission rate. Both BACT and LAER are determined for the source by the permitting authority on a case-by-case basis.

The PSD program is a pollution prevention program, or a statutory embodiment of the "precautionary principle". The social and legal relationships that are specified in part C, sections 160 and 165 of the Act state: (1) that it is up to the state to protect public health and welfare from any actual or potential adverse effect from air pollution, *notwithstanding attainment and maintenance of all national ambient air quality standards*; (2) that the Act has as an additional purpose: the preservation, protection, and enhancement of air quality in national parks, etc.; (3) that air quality managers are to ensure that economic growth occurs in a manner consistent with the preservation of clean air resources; (4) that one state's emissions should not interfere with any other state's PSD plan; and, (5) that no decision to permit increased air pollution in any area is to be made without evaluation of all the consequences of such a decision or without adequate procedural opportunities for informed public participation in the decision-making process. The environmental quality objectives in the PSD program include not only the maintenance of air quality to meet the NAAQS, but also the minimization of any other effect a proposed source's emissions might have on, for example, soils, vegetation and visibility. With respect to national parks, there is a special responsibility to ensure that proposed sources do not adversely affect any air quality-related values, such as visibility, stream chemistry, vegetation, and cultural resources and other factors that are important to the park's integrity. The BACT provision is thus an independent requirement to apply the best technology to each proposed source under a broader PSD authority to prevent all significant impacts of any new source.

In nonattainment areas, proposed new sources must not only meet LAER but also have to obtain offsets at a greater than one-to-one ratio for its projected emissions. The offset requirement was initiated in 1976 to provide necessary relief in nonattainment areas where a nominal restriction on any increase in nonattainment pollution could have meant a moratorium on the construction or expansion of any major new facility. Under the new law (the 1990 Clean Air Act amendments), the offset ratio varies depending upon the severity of the nonattainment problem. Also in nonattainment areas, *existing* sources may be required to use RACT (reasonably available control technology), an exception to the general rule under the Act of applying BAT requirements to new sources only. This mention of offsets brings us to the subject of emissions trading and the early experiences with it under the Act. We return to a discussion of the difference in the Act's treatment of new and existing sources in Section 6.8.

6.7 A Brief History of Emissions Trading since 1976

If strictly applied, the pre-1977 CAA would have prohibited the construction of new major sources of any nonattainment pollutant. To avoid the economic impact that situation would have fostered, and the backlash against air quality requirements, the EPA developed the offset policy in 1976, creating the first prong of what would become a four-prong emissions trading policy. The offset policy allowed proposed new sources to obtain offsetting emission reductions from another source in the same geographic area so that, despite the emissions from a new source, overall emissions would decrease. The term emission reduction credit (ERC) came into being, representing the amount of "extra" emission reductions a source can achieve beyond what it is required to achieve under the law. ERCs became a sort of currency, allowing transactions between two or more sources so that the overcontrol by the first source could be used or credited toward the compliance obligations of another.

The emissions trading policy was developed over ten years, with the final Emissions Trading Policy Statement issued in 1986. The policy includes specific aspects of emissions trading: offsets, netting, bubbles and banking. The offset program was discussed; it is the second most widely used of the emissions trading policy programs after netting. Netting, the most frequently used provision under the emissions trading policy, is used in both attainment and nonattainment areas to avoid triggering new source review. As discussed above in the section on the technology requirement, all major new or modified sources must undergo a preconstruction review and permitting process. A source is considered major if it will emit more than a specified number of tons per year of a listed pollutant, including the six pollutants for which NAAQS exist and certain toxic air pollutants. Netting allows a company to avoid triggering new source review by letting the source average its emission increases with enforceable reductions at existing units on the same property.

The bubble concept, unlike netting and offsetting, is at least theoretically based on ambient considerations, the idea being that two or more sources of emissions can be regarded as existing under an imaginary dome, or bubble, under which the effect of any of the bubbled sources' emissions on ambient air quality would be the same. Finally, the Emissions Trading Policy Statement allows states to create a "bank" of emission reductions that have been certified as ERCs. Banked ERCs can theoretically be used for netting, offsetting or bubbling. To be eligible as an ERC an emission reduction must meet four criteria:

1. Permanent: the reduction cannot be seasonal, periodic, or temporary.
2. Surplus: the reduction must not be required by any other existing requirement and it must result in an actual decrease in emissions.
3. Enforceable: the reduction and its method of creation must be legally binding and enforceable.
4. Quantifiable: the amount of the reduction must be able to be quantified by an accepted method or calculation technique.

This brief outline of the major components of the emissions trading program does not begin to represent the complexity of the program or the degree of controversy that surrounded its development. A good part of the debate over the pre-1990 emissions trading program must be understood in the context of the weakness and flaws of the SIP process, especially given the persistence of the nonattainment problem.

It would be difficult to overstate the controversy that surrounded the use of emissions trading in the USA in the years between the 1977 Clean Air Act amendments and the 1990 amendments. At the heart of the disagreement was the mismatch between emissions trading proposals and the underlying statutory goals and specific requirements which were demonstrably not up to the task of achieving attainment for all the environmental quality objectives in all areas. Emissions trading, by definition, embraces the idea that there are "extra" emission reductions that can, and should, be achieved. But this was during a decade when the underlying goals of the CAA were not being met. Despite some clear successes such as the lead phase-out program, there were some serious failures, and certain problems of public concern were not even being addressed, such as visibility degradation and acid rain. There was thus widespread disagreement over what would constitute "extra" emission reductions that could, and should, be bubbled, netted, used as offsets or banked, instead of being used directly for environmental improvement. "Extra" is a matter of opinion with legal, economic and social dimensions. Furthermore, the fact that emissions trading was being promoted during the Reagan Administration, which was openly hostile to regulation in the first instance, did not increase confidence in these proposals. Regulatory reform, or specifically, the economic reform of environmental regulation, an idea that began under the Carter Administration in the late 1970s, became regulatory relief in the 1980s. It took until 1988 to get reform back onto a constructive track.

One of the most contentious issues related to economic instruments during the decade from the mid-1970s to 1986 when the Emissions Trading Policy Statement was finalized was the problem of the "baseline". To generalize, most emissions trading advocates promoted rules that would allow a source to trade on the basis of its "allowable" emission limit (the SIP limit referred to above), arguing that because the limits had been approved by EPA as being adequate to attain or maintain the NAAQS, they should also be accepted as the baseline for bubbles. In other words, they would have interpreted the social and legal relationship established under the Act as resting solely on what a state promulgated in the SIP, regardless of whether the SIP could in reality achieve the health-protecting objectives of the law.

Again, to simplify, the Air Office at EPA, in recognition of the weakness in real-world SIP modeling and emissions trading, advocated the use of "actual" baselines, a requirement that admittedly was more stringent for sources that wanted to engage in emissions trading than for others. The exception the Office of Air and Radiation (OAR) allowed to the use of "actual" baselines would be if sources undertook detailed air quality modeling based on their "allowable" baselines to demonstrate their maintenance of the NAAQS, PSD increments (another EQO) and, in nonattainment areas, reasonable further progress toward attainment. The OAR, it can be said, insisted that the purpose of the social and legal relationships between emitter and the regulators was the one described earlier in this chapter: to do everything possible to meet the health-protecting goals. Of course, the emissions trading debate was much more complicated than described here, and interested readers have a large literature to consult. The real problem was that the emissions trading proposals were being grafted onto a statutory and regulatory framework that simply couldn't bear the weight.

In 1990, the amendments again revised the law to try to find a way to accomplish the Act's primary objectives, and some of the earlier problems in conceiving and implementing emissions trading and related instruments may have been ameliorated by the 1990 amendments' specifications for nonattainment areas. In general, the new law gives the EPA and states the authority to develop better, more complete SIPs that cover more sources, allow for better accountability, and require better inventories and other measurement tools. Specifically, the authorities in nonattainment areas must promulgate revisions to its SIP (whether or not they intend to pursue an economic incentive program) that include the following: an inventory of actual emissions from all sources; the adoption of all reasonably available control measures, including the technology-based RACT requirements for existing sources; provisions to ensure that reasonable further progress is made toward attainment; a special permit program for new and modified sources; and any other measure that may be necessary to ensure attainment. In short, the new nonattainment provisions provide for much further coverage of emissions and activities in nonattainment areas.

This reinforced institutional structure, it is hoped, will support creative or innovative strategies that may involve economic instruments, with less risk of erosion or delay in progress toward attainment. To that end, the EPA has proposed a new set of rules for the development of economic incentive programs that may be adopted by states for certain nonattainment areas. Under certain circumstances, states may be obligated to adopt such a program. The proposal identifies key program provisions which must be included to ensure that it will not interfere with other requirements of the Act; and that emission reductions credited to the program will be: quantifiable; consistent with SIP attainment and reasonable further progress demonstrations; surplus to reductions required by, and credited to other implementation plan provisions (to avoid double counting); enforceable at state and federal levels; and permanent within the time frame specified by the program.

The proposal specifies that an economic incentive program should include certain elements:

- A clearly defined purpose and goal, and an incentive mechanism that can be rationally related to accomplishing the goals.

- A clearly defined scope, which identifies affected sources and ensures that the program will not interfere with any other applicable federal regulatory requirements.
- A program baseline from which projected program results (e.g., quantifiable emission reductions) can be determined.
- Credible, workable, replicable procedures for quantifying emissions and/or emission-related parameters.
- Source requirements, including those for monitoring, record keeping, and reporting, that are consistent with specified quantification procedures and allow for compliance certification and enforcement.
- Requirements for projecting program results and for dealing with uncertainty.
- An implementation schedule, administrative system, and enforcement provisions for ensuring federal and state enforceability of the program.

In addition, in certain cases, the program must include audit procedures to evaluate program implementation and track results, and reconciliation procedures to trigger corrective actions or contingency measures to make up any shortfall between projected emission reductions and those actually achieved in practice.

The above list of essential elements may strike some readers as overly regulatory, especially if they labor under the impression that incentive-based regulation is some sort of antithesis to "command-and-control". Economic incentive programs are meant to provide individual sources with the flexibility to choose and vary their specific compliance strategies within a commitment to meet specific overall goals. These programs are meant to lead to faster and/or more cost-effective attainment of air quality goals. Which elements in this list of general requirements are unreasonable or superfluous? Is it too much to ask the drafter of an incentive program to clearly specify goals and objectives and to relate the specifics of the proposed program to the goals. Is it onerous to require someone getting credit under a SIP to keep track of their performance through monitoring and reporting requirements? If a bright idea does not work, should those who would have benefited from its success just walk away when it fails, or should there be contingency plans in place?

The economic incentives program proposal differentiates among three broad categories of program: emission limiting, market response, and directionally sound. Emission-limiting strategies directly specify limits on total mass emissions, emission-related parameters, or levels of emission reductions relative to a program baseline that are required to be met by affected sources. A marketable permit program such as the acid rain allowance system would fit under this category, although the acid rain program itself predates this rule and exists under its own statutory authority. The second category, market response, creates incentives for affected sources to reduce emissions without directly mandating emission-related requirements for individual sources. An emission fee system would fit under this category. The consequence of such a program is that actual emissions from affected sources will be unpredictable and may differ from projected emissions used in the pre-program implementation plan. This outcome uncertainty must be accounted for in the design of the program. Finally, economic incentive programs which can not yield quantifiable emission reductions creditable toward reasonable further progress or attainment demonstrations (because they lack one or more of the basic program elements listed above) may still be included in a SIP but may not be credited. Certain transportation control measures or labeling programs may fall into this category, depending upon the program design.

6.8 New Source Bias Arguments and Emissions Trading

Outside the debate over the use of economic incentives programs in nonattainment areas there has been a separate debate over the difference between the treatment of new and existing sources – in shorthand, the "new source bias" of the Clean Air Act. Although related to the broad emissions trading debate, it is really a narrower issue, and the acid rain control program can be viewed as the last word (so far) on the subject.

The underlying logic of the pre-1990 Clean Air Act as far as stationary sources are concerned is that air quality progress could be achieved with least economic disruption if the nation focused most of its pollution control effort on new sources, allowing old, relatively uncontrolled sources to live out their useful lives and retire, taking their pollution with them. The CAA focuses its pollution control requirements on new plants based on certain assumptions about capital stock turnover; i.e., that old plants would be retired and replaced with new ones that would be cleaner and cleaner over time. This turns out to have been a very flawed assumption at least for fossil fuel-fired power plants. In 1995, 25% of the installed electricity capacity (34% of the fossil fuel-fired steam capacity) will be more than 30 years old, and reaching the end of its nominal design life (Baylor, 1990). According to an industry analyst: "there is no way a significant percentage of existing capacity can be retired during the next twenty years. It is irreplaceable in the short term. ... With little new capacity and retirement of old capacity planned, life extension is inevitable" (Smock, 1989). Life extension is the term used to connote the retention in service of a generating unit beyond its original design or economic lifetime. In 1990 over two-thirds of acid rain-causing pollution from power plants was emitted by plants constructed before 1970.

Some critics of the Clean Air Act, who have also tended to be emissions trading advocates, purport to demonstrate that the law's new source bias – that is, a bias against new sources – has contributed to more overall emissions by increasing the relative marginal cost of a new facility compared to an existing one, leading firms to maintain their old, higher emitting facilities longer. *Clean Coal/Dirty Air* is one oft-cited work (Ackerman and Hassler, 1981), which focuses on the new source performance standards provisions of the 1977 CAA amendments. Robert Crandall, in *Controlling Industrial Pollution* (1983) provided more analysis, however, arguing that the new source standards "are so stringent that the incremental costs of control seem to be higher than for retrofitted existing sources. This situation can only retard the development of new sources, increase the cost of pollution control, and slow the cleanup of the environment." Convinced that the new source bias was politically motivated to keep industry from relocating to the south and west, and not to protect air quality, or manage the increase in pollution attending economic development (or even both), Crandall recommended: "All new source performance standards should be abolished. New sources should be required to buy pollution rights (offsets or transferable emission credits) for any emissions that they generate, but their emissions should not be prescribed by government" (Crandall, 1983, p. 169). He also argued that "the prevention of significant deterioration (PSD) policy should be altered to *allow* industry to migrate to the growing sections of the country, particularly if these areas have cleaner and more healthful air than the declining northern and northeastern states" (Crandall, 1983, pp. 169–170, emphasis added).

It is true that the 1977 new source performance standards (NSPS) required all new combustion sources to install pollution control equipment and to achieve a given percentage reduction in emissions, regardless of the sulfur content of the coal. It is widely recognized that this served to promote the relative attractiveness of eastern high-sulfur coal over western low-sulfur coal. The 1970 CAA amendments had done just the opposite: western and eastern low-sulfur coals had gained some competitive advantage at the expense of eastern high-sulfur coal. The third round of this interregional competition was concluded during the Congressional debate over the acid rain control amendments.

But what is not widely acknowledged in the new source bias discussions are the beneficial environmental consequences of the percentage reduction requirements. For example, there are 10,203 MW of scrubbed NSPS capacity in the 11 western states, out of a total capacity of 56,570 MW. Assume that on average these scrubbed NSPS units emit at an annual average rate of 0.31 lb/MBtu (lb/MBtu: pounds per million British thermal units); with an average heat rate of 10,000 Btu/kWh (Btu/kWh: British thermal units per kilowatt-hour), and a 70% plant capacity factor, these units emit just under 100,000 tons of SO_2 per year. But had these units not been scrubbed, they would have emitted over 300,000 tons of SO_2 per year – over three times as much on an annual basis. Also, the 0.31 lb/MBtu emission rate assumed in this example is conservative, i.e., it overestimates actual emission rates at many NSPS units in the west. Several western states require their new sources to emit at much lower rates (0.08–0.20 lb/MBtu are more common) either through state-specified NSPS limits or case-by-case determinations of BACTs under the PSD program. Thus the *increase* in annual emissions that we might have seen without the new source standards would have been even greater than estimated here.

The new source bias arguments focus on the fact that the marginal cost of control of scrubbing western rather than eastern coal is three to four times higher on a cost per ton basis, leading authors such as Crandall to conclude that the investment in scrubbers for western power plants is inefficient. By assumption, they concluded no marginal benefits. But the marginal environmental benefits per ton of sulfur avoided are much higher in the west because of the very low levels of background pollution. Had the 1970 NSPS stood for the past 12 years, roughly 300,000 more tons per year of sulfur would have been emitted into some of the most pristine air, over some of the most esteemed natural resources (the national parks, wilderness areas) in the world. Consider, for example, the recent agreement reached by the owners and operators of one power plant in Arizona, the Navajo Generating Station, located within the golden circle of class I areas in the southwest, and several environmentalist groups concerned about the plant's impact on the Grand Canyon and other national parks. This one plant, constructed before the 1977 new source standards were promulgated, is now finally going to clean up to new source levels. By the late 1990s, the Navajo plant will reduce its sulfur dioxide emissions by roughly 90,000 tons per year at an estimated annualized cost of $90 million, with estimated net present value of benefits in the range of \$1.7–\$2.6 billion (\$250 million annualized benefits).

Should we accept Crandall's claims that the 1977 NSPS must have induced utilities companies to "postpone the replacement of older, obsolete units, thereby saddling consumers with higher generation costs?"(Crandall, 1983, p. 40) Especially in the west, which has seen such an increase in population, electricity demand growth and power generation, and given all the other relevant factors involved in power plant siting and construction, it is not likely that the marginal costs of new source requirements have in fact delayed the timely retirement and replacement of older sources. In fact, the relative likelihood of power plant retirement is continually falling. EPRI and the power generation industry see no real reason ever to retire power plants. In the absence of pollution controls, a new unit would cost on the order of \$1000–2000 per kW, while to extend the life of an existing unit would cost \$200 per kW at most. While different pollution control requirements increase that differential, there is already such a considerable spread that the incremental impact on the "build-new-or-life-extend" questions facing all utility managers is speculative.

Furthermore, in the ten years since the publication of Crandall's book, there is little to suggest that his concern over delayed economic growth in the Sunbelt due to new source standards, particularly for power plants, was warranted. Florida and California were among the fastest growing states; even with the new source standards in place, the EPA's 1989 base case projected that Florida's SO_2 emissions would increase from 501,000 tons per year in 1985 to 933,000 tons per year in 2000. And today the state of Virginia has PSD permits pending for roughly two dozen new power plants. Without the strict new source requirements under NSPS and PSD, these new facilities, located near the Shenandoah National Park and the James River Face Wilderness Area, would rapidly degrade air quality in these class I areas. Considerable controversy exists over these proposed plants, even with the strict controls proposed for them given their location near these class I areas and the cumulative impact of the new and existing facilities on areas that have already been considerably impaired.

Anyone with an elementary knowledge of economics feels the intuitive appeal of the new source bias arguments: unless the price elasticity of demand for new cars or new product factors (like capital investment in new steel or power plants) is zero, an increase in the cost of new goods must have some effect on the demand. But the empirical evidence for the magnitude of this effect, given all other political and economic influences on new build, new buy, and location decisions is much too weak to support the most popular reform recommendation: to weaken the new source standards. The economic efficiency arguments against the Clean Air Act's BAT-type requirements (NSPS and BACT) would have been more compelling had the critics also worked to tighten the requirements for existing sources. As it was, the new source bias arguments mostly surfaced in support for regulatory relief – weakening of the new source requirements as a goal in and of itself, without a corresponding effort to tighten the requirements on the sources of most of the pollution.

6.9 Acid Rain Control as Regulatory Reform

In contrast, the acid rain control program established under the 1990 amendments is a policy of regulatory reform: it reforms regulation by correcting the serious mistake in the underlying Act (where existing sources are left relatively uncontrolled and stringent new source standards are relied upon to gradually improve and maintain air quality).

The acid rain policy deals with the new source bias problem, not by relaxing standards for new units, but by strengthening the requirements on existing ones. The new source standards (as well as the NAAQS-related provisions) are retained under the mixed portfolio approach to air quality management. In this sense it is clearly regulatory reform but not regulatory relief. It is also regulatory reform in that it includes an allowance trading system; this is a specific example of an emissions trading scheme that has often been advocated by proponents of economic reform of environmental regulation in order to reduce the cost of environmental protection.

In brief, the acid rain control program requires a 10 million ton reduction in sulfur dioxide emissions relative to 1980 levels. The acid rain title of the 1990 CAA amendments concentrates its regulatory requirements on the electric utility industry, although roughly 1.5 of the 10 million tons of reduction are expected to come from non-utility sources of SO_2. The basic structure of the program is simple. Reductions in SO_2 emissions will be achieved in two phases. Phase I begins on 1 January 1995, and is designed to achieve a 5 million ton reduction in SO_2 emissions relative to 1980 levels. Phase II begins on 1 January 2000, and will complete the title's 10 million ton reduction goal. The remaining allowable annual emissions for electric utilities is 8.9 million tons. The amendments then impose a cap on total allowable SO_2 emissions from this sector at 8.9 million tons per year, beginning in the year 2000.

Each electric utility unit in the country above 25 MW_e in size, is allocated a portion of the 8.9 million tons of allowable SO_2 emissions through a new regulatory instrument called an emissions allowance. Each allowance is an authorization to emit one ton of SO_2. Thus, when fully implemented, the acid rain program will issue 8.9 million allowances per year. The statutory formulas that allocate allowances serve as default limits on allowable annual emissions of SO_2 for each unit. That is, beginning the moment the amendments were signed into law by the US President, each electric utility unit was under the affirmative obligation to emit no more SO_2 than the amount specified for it in the statute (to become operable on the deadlines described above). Had there been delay or stalemate over the implementing regulations, the industry would have faced a much more rigid program than the one made operable by the promulgation of the emissions allowance trading system regulations.

Under this set of regulations, the EPA will operate an emissions tracking system, to keep track of actual performance on a unit-by-unit basis and an allowance tracking system, to keep track of where allowances have been transferred. The operative provision is the prohibition against emitting SO_2 (on an annual basis) in excess of the number of allowances held by the unit. The number of allowances held by any unit would equal the number initially allocated by statutory formula plus any procured, minus any sold or otherwise disposed of. The acid rain title does not call for any review of or restriction on proposed allowance trades. Essentially, the acid rain program itself is spatially indifferent; it is based on emissions only. However, all the other regulatory requirements continue to apply. Thus, regardless of how many allowances a unit may hold, it cannot emit SO_2 at a rate greater than its SIP limit, or its limit under the NSPS and PSD provisions.

Now that the CAA includes provisions covering local and regional air quality (at least near class I areas), and acid rain provisions aimed at long-range transported and transformed air pollution, some still question whether the public should be concerned about where exactly the acid rain control reductions are achieved and where a plant might be allowed to increase its emissions. Others formulate the questions differently: given the full set of air quality control requirements that constrain a utility's emissions, will any allowances ever be traded? The answer to the second question is yes, empirically yes. The industry, aided by third parties serving as brokers, gives every indication of having figured out how to use the flexibility of the acid rain program to save compliance costs and to optimize their system compliance strategy.

As for the first question, on whether or when the public should be concerned about the environmental consequences of an allowance trade, consider the following three scenarios. First, if the underlying SIP governing a plant's emissions is suspect or is known to be flawed, then the state or others may apply pressure to at least examine the relevant parts or the SIP, if not amend a faulty permit, before final acid rain compliance plans are made to ensure that the plant really is not causing or contributing to a violation of the SO_2 or PM-10 (particulate matter 10 μm or less in size) NAAQS or PSD increment. A utility can readily check for itself whether its emissions may be suspected of causing or contributing to a NAAQS violation or increment consumption using the EPA's air quality modeling guidelines.

Second, although the EPA has not revised the SO_2 NAAQS to include a very short term averaging time (to prevent brief, but exceedingly high concentrations of SO_2 in the ambient air near a source), the environmental community is pressuring them to do so, and it is possible that a state could adopt a short-term standard on its own. Even without a change of standard, a screening model (of the ambient air quality impact) can be used to predict whether any particular plant might cause such exceedances; if so, local interest groups might profitably emphasize this fact rather than dwell on changes in annual emissions allowed under the acid rain program which may not cause or contribute to any residual air quality problem, particularly not a short-term, local one. Again, utilities can arm themselves with this information rather than wait and be surprised.

Finally, if a plant's emissions will cause or contribute to visibility impairment or other air quality-related impacts in a class I area, regional environmentalist groups may profitably take legal action to force a utility or state to take such concerns into account. They may also continue to pressure the EPA to put into effect the long-awaited regional haze regulations and to support the federal land managers' attempts to protect their class I areas from air pollution. Utilities can take into account the demand for less SO_2 and NO_x exposure in the class I areas and make excess emission reductions at nearby plants, and then use the allowance market to offset whatever incremental costs they might incur.

No one has argued that the acid rain policy is capable of protecting the most sensitive resources from sulfates and nitrogen species. The allowance market cannot substitute for local and regional air quality protection measures. Nor, however, without any specific air quality-related concerns, should resistance to allowance trading – whether buying or selling – be allowed to interfere with allowance transactions. Such interference can only lead to considerably higher acid rain compliance costs, which ultimately will be paid for by electric utility ratepayers.

6.10 Discussion: Between Promise and Practicality

For over two decades the environmental economics literature has presented marketable permits and emissions taxes as superior alternatives to conventional regulation based on the results of their optimization modeling. But how good is the link between policy models and real policy problems? What about the institutional environment – how, exactly, will trades be negotiated and made? The gap between model processes and real procedures distorts the expectations of policy analysts and decision makers. Some of the mathematical models used in air pollution policy analysis essentially solve for a static, partial equilibrium under conditions that in the real world would reflect simultaneous trading transactions among fully informed individuals at no cost. In reality, traders have very limited information, incur real transaction costs, and make bilateral, sequential trades. The "expectations gap", that is, the disappointing cost savings realized under the bubble policy and other aspects of emissions trading compared to the theoretical savings predicted by trading advocates, has as much to do with unrealistic expectations as with policy-important constraints like nondegradation or anti-backsliding requirements which prohibit a participant in a trade from actually increasing emissions above existing levels (Atkinson and Tietenberg, 1991).

The US acid rain program is fundamentally different from the emissions trading program and, while it is just getting under way, the expectation is that the careful attention to institutional structure in the authorizing statute will motivate and support a fairly fluid allowance market and still ensure that the program's goals are met and that local and regional environmental protection is not jeopardized. While the acid rain program *is* premised on a positive answer to the question posed above about whether a more flexible regulatory structure could lead to more efficient and cost-effective controls, that answer is in turn based on the retention of the local and regional air quality protection provisions of the underlying Clean Air Act, and upon the continued authority of the technology-based standards. In addition, the acid rain program has very clear goals and a structure comprising rigorous administrative control mechanisms – from the permanent cap on allowable SO_2 emissions and an unassailable baseline, to the allowance allocation rules, to the emissions monitoring, recording and reporting requirements, and to the excess emission penalties set far above the marginal cost of control. In considering the qualities of the acid rain program, or any other regulatory regime, the administrative structure cannot be separated from the primary objectives: you have to be able to "get there from here".

The qualities that support a functional institutional structure, such as those just described, do not inhere in using incentive-based strategies; they result from self-conscious visible decision to build such qualities into whatever program is chosen. The emissions trading provisions may have lowered the cost to firms of compliance with the Clean Air Act (a not unimportant factor), but there is very little hard evidence to support that conclusion. There is even less evidence they have provided any increase in environmental benefits, or encouraged innovation and technical change, or increased economic efficiency beyond what would have occurred under a comparable "command-and-control" program with appropriately designed institutional structures. On technological change, proponents of emissions trading tend to make claims such as the following: "incentive mechanisms provide a greater stimulus for innovation and technical change in pollution control than does a direct regulatory approach" (US EPA, 1992, p. 2-1) The problem with such categorical statements is that technological change depends on a large number of factors, and it is far from clear that anyone knows what affects the rate of technological change and what the rate of change means for the economy as a whole (Kemp and Soete, 1992). An important related factor not usually evident in the literature is what might be termed the "effective demand" for cleaner technology. For example, trading SO_2 emissions at or near the pre-acid rain control level of roughly 18 million tons of SO_2 per year for the utility sector, would not have promoted technological change in the electric utility industry. At that overall level of emissions (the 1985 level) it simply would be too easy to limit emissions with existing technologies for emissions trading to have had much effect on technological change. But the effective demand for cleaner power plant technologies, fuels and management strategies might increase significantly when the allowable limit is halved to approximately 9 million tons per year. This stringent goal, coupled with the combination of flexibility and accountability embedded in the acid rain legislation and implementing regulations, represent the state of the art in incentive-based regulation. Now we just have to be patient to see whether it works.

Note

The views expressed in this chapter are those of the author. They do not necessarily represent the views of the Environmental Protection Agency, the Department of State, or any other agency of the US government.

References

Ackerman, B.A., and Hassler, W.T., 1981, *Clean Coal/Dirty Air*, Yale University Press, New Haven, CT, USA.

Atkinson, S., and Tietenberg, T., 1991, Market Failure in Incentive-Based Regulation: The Case of Emissions Trading, *Journal of Environmental Economics and Management* **21**:17–31.

Baylor, J.S., 1990, Acid Rain Impacts on Utility Plans for Plant Life Extension, *Public Utilities Fortnightly*, March 1, p. 24.

Clean Air Deskbook, 1992, Environmental Law Institute, Washington, DC, USA.

CRS (Congressional Research Service), 1974, *The Legislative History of the Clean Air Amendments of 1970*, Vols. 1 and 2, Government Printing Office, Washington, DC, USA.

Crandall, R.W., 1983, *Controlling Industrial Pollution*, The Brookings Institution, Washington, DC, USA.

Hawkens, D., 1981, Testimony, (no title) published in "Hearings before the Committee on Environment and Public Works, United States Senate, 97th Congress, First Session, April 8, 9, May 20, and June 2, 1981, Part 1, US Government Printing Office, Washington, DC, USA.

Kemp, R., and Soete, L., 1992, *The Greening of Technological Progress: An Evolutionary Perspective*, Maastricht Economic Research Institute on Innovation and Technology (MERIT), *Futures* **24**(5):437–457.

Kete, N., 1994, Environmental Policy Instruments for Market and Mixed-Market Economies, *Utility Policy Journal* **4**(1):5–18.

Smock, R., 1989, Power Plant Life Extension Trend Takes New Direction, *Power Engineering* **93**(2):21–27.

US EPA (Environmental Protection Agency), 1992, *The US Experience with Economic Incentives to Control Environmental Pollution*, EPA-230-R-92-001, Washington, DC, USA.

Chapter 7

Economic Instruments for Air Pollution Control in Estonia

Tiit Kallaste
Stockholm Environmental Institute
Tallinn, Estonia

Abstract

This chapter describes the experience of Estonia in the implementation of economic instruments for air pollution control. Market economy principles in environmental policy in Estonia have not been in operation for very long; the systematically organized taxation of environmental pollution actually started only in 1991. The most important market forces applied in Estonia today are the so-called economic instruments and resource pricing. Several economic instruments for the control of air pollution have been introduced to complement other means, such as direct regulation, particularly emission charges, subsidies and noncompliance fees, and are characterized in detail. Also the Estonian Environmental Fund has been the basis for the implementation of economic instruments to control environmental pollution and for resource pricing. The Fund serves as an institution for collecting revenues for environmental protection purposes. On the basis of Estonia's experiences in the last three years with economic instruments under the very difficult conditions of the transition to a market-oriented economy, one can conclude that the revenue-raising character of economic instruments is dominant since emission charge rates have been set relatively too low to provide effective economic incentives to polluters to reduce their emissions.

Key words: economic instruments, emission charges, toxicity of pollutants, maximum permissible concentrations, revenue raising, enforcement incentives, subsidies, noncompliance fees, environmental protection fund.

7.1 Introduction

The intensive and large-scale use of natural resources, particularly oil shale and phosphates, caused significant damage to the natural environment in Estonia during the period of the centrally planned economy, particularly in the north and northeast of the country. All components of the environment have been seriously damaged by pollution in this area. Air pollution has been particularly acute, and has had negative effects on human health in the most polluted areas in the northeast (Estonian Nature Management Scientific Information Centre, 1990; Ministry of the Environment, 1991; National Board of Waters and the Environment, 1992; Kallaste, 1991), although the air pollution situation is somewhat better in other parts of Estonia (Kallaste *et al.*, 1992).

During the postwar socialist period, the possibilities for economic control over polluting sectors of the economy in Estonia were very restricted, since they were centralized by All-Union ministries. More than 90% of all Estonian industries were controlled by the central government in Moscow; Republican (i.e., Estonian) authorities had very little influence on the planning of the structure of local production. As a result, there was a lack of environment-oriented investments, particularly for air pollution abatement, so that air pollution reached critical levels in many places.

Many international environmental agreements on SO_2 and NO_x emissions are in force nowadays. For that reason investments are needed to buy or build modern purification equipment or low-waste technologies – the faster, the better. Until recently polluting enterprises were not concerned about the high levels of air pollution, as there were no real economic incentives for them to reduce it. Now changes are needed to update cleaning facilities and technologies. Environment-oriented investments not made during the last decades need to be made in the near future. The major restricting factor is the shortage of hard currency for the purchase of modern production and purification technology. Experts estimate that to clean up the sulfur emissions from oil shale power plants in the northeast of Estonia, for example, will cost at least $1 billion.

To control the pollution, several market-oriented environmental policy instruments for economic management have been introduced in Estonia step by step since the early 1970s. Enforcement incentives such as noncompliance fees or fines have been imposed for air and water pollution. However, the principles of taxation of the environment were not based on classical economic theory, but on socialist political economy. Taxes were introduced on the basis of expert assessments, in view of the need for immediate restrictions on pollution. Tax rates were calculated on the basis of finances available (or necessary) for pollution abatement activities during each year, or on health criteria. Estonia was one of the first republics of the former Soviet Union to implement economic instruments to control environmental pollution.

7.2 Economic Instruments in Estonian Environmental Policy

7.2.1 Introduction

Economic instruments are environmental policy tools that take advantage of market principles to achieve the goals of environmental improvement. They are often used as substitutes for or as complements to other policy instruments such as regulations and cooperative agreements with industry (OECD, 1991, 1992). Policy instruments in general can be classified into three main categories (Bye *et al.*, 1990): economic instruments, regulatory instruments, and information or moral suasion.

Economic instruments offer a number of inherent advantages over regulations and moral suasion, including greater cost-effectiveness, permanent incentives to reduce pollution, greater flexibility in application, and the provision of a continuous source of revenue. One basic objective of economic instruments is to ensure appropriate pricing of environmental resources in order to promote their efficient use and allocation. The cost-saving potential is one of the major characteristics of economic instruments. According to an OECD definition,

> ... economic instruments affect the costs and benefits of alternative actions open to economic agents, with the effect of influencing behavior in a way that is favorable to the environment. Economic instruments typically involve either a financial transfer between polluters and the community (e.g., various taxes and charges, financial assistance, emission reduction credits, user charges for services, product charges), or the actual creation of new market systems e.g., tradable emission permits. [OECD, 1991, p. 10–11]

According to the OECD (Opschoor and Vos, 1989; Nicolaisen and Hoeller, 1990), economic instruments can be classified as follows:

1. *Charge systems:* effluent charges, user charges, product charges, administrative charges and tax differentiation.

2. *Subsidies:* financial aid for installing new technologies, and subsidies for environmental R&D expenditure; grants, soft loans, and tax allowances can be differentiated here.
3. *Deposit–refund systems:* combine charges and subsidies so as to provide incentives to return pollutants for recycling.
4. *Market creation or redefining property rights:* tradable emission permits, market intervention, and liability insurance legislation.
5. *Enforcement incentives:* noncompliance fees, performance bonds; these are inseparable from regulatory measures.

Two of these five economic instruments are not currently in use in Estonia; deposit–refund systems and market creation, although initial steps toward introducing liability insurance were made in 1992. Subsidies and enforcement incentives have been practiced for some time, combined with various regulatory instruments. In the following, some characteristic examples of the implementation over the last few years are presented (Kallaste, 1993).

7.2.2 Charges

Charges are new instruments for pollution control in Estonia. Three types exist: emission charges, administrative charges, and tax differentiation.

Emission charges are, in principle, fixed taxes (so-called regular confirmed emission charges) on direct releases into the air. Ambient concentrations must not violate health standards, or daily maximum permissible concentrations (MPCs; see *Table 7.1*). Charges have been set for SO_2, NO_x, CO, nontoxic dust, oil-shale fly ash, soot, etc. In practice they are levied on up to about 50 pollutants contained in stack emissions. In general, the charges are calculated on the basis of two extreme standards: the working area and MPCs within a 24-hour period. Sulfur dioxide is used as the basis.

Emission charges for air pollution control were first implemented in January 1991. Regular emission charges are imposed by executive bodies of local governments on the basis of the entrepreneurship contracts (Government of Estonia, 1990). All polluters are supposed to pay these so-called confirmed regular emission charges. According to government regulations, however, not all polluters are obliged to pay the charges; nonpoint polluters (motor transport, sea and river transport), as well as households and institutions financed from the state budget, are exempted.

Local authorities have the right to propose regular emission charge rates, and to use the revenues themselves, for investment, environmental research, development, etc. They have the right to increase or to decrease the charge rates established by the government by 20%, according to local conditions and the economic situation of an enterprise (Government of Estonia, 1992). In special cases the emission charges may be applied gradually, starting from relatively low rates. In this way it works as a kind of subsidy in the form of tax allowances.

Administrative charges are not widely used, although there are many areas where control and authorization fees are to be paid to the regulatory authority. For example, the annual technical inspections of all vehicle exhaust gases with respect to CO content could be classified as an administrative charge; this has been practiced in Estonia for more than 10 years. Administrative charges are also used in the case of emission permits for polluting enterprises.

A differentiation of the tax on gasoline was introduced several years ago to encourage the use of lead-free fuel, but failed due to problems with obtaining regular supplies from the former Soviet Union. At present, the fuel supply is stable and various types of fuel are continuously available. The system of tax differentiation should be reintroduced now. There are as yet no examples of implementation of tax differentiation in other fields.

7.2.3 Subsidies

Subsidies have been used in combination with regulatory instruments for a long time. Most subsidies are in the form of grants or tax allowances, particularly for environmental investments. Grants from the state budget for air pollution abatement were the main form of subsidy during the period when Estonia had a centrally planned economy. In fact, most of the environment-oriented investments were financed from the state budget. The distribution of investments in different spheres was traditionally made on the basis of expert assessments, but the shares of investments in water, air, waste management, etc, have been uneven. The share of investments in air pollution abatement has always been less than 10% of total investments in environmental protection.

The criteria for allocating grants were quite different from those used in OECD countries. Environmental investments were financed from centralized funds of the ministries or from the state budget. The centralization and redistribution of profits to enterprises of the ministries was common, so that the subjective investment decisions of the central government in Moscow or in Estonia prevailed. These decisions were not always based on comparative analyses of the cost-effectiveness of alternative investment projects. On the whole, the centralized environmental investment policy was not very efficient.

Today, tax allowances in the form of emission charge exemptions are also granted if a polluter has voluntarily installed gas cleaning equipment or has purchased environmentally friendly technology. In fact, it depends on the willingness of an enterprise to introduce less polluting technology, or purification equipment or waste treatment systems. The borderline for obtaining a tax allowance is set so that the enterprise itself has to achieve a reduction of at least 25% of emissions into the air. The polluter is exempted from the emission charge by 150% of the investment cost for environmental improvement achieved during the fiscal year.

7.2.4 Enforcement incentives

The noncompliance fee is a kind of enforcement incentive, which was introduced in connection with emission charges in 1991. If a polluter exceeds the confirmed regular emission limits, a much higher tax is levied according to the toxicity of the pollutant. Four toxicity classes are used, based on the health hazards of the pollutants. The tax rate is determined on the basis of expert estimates and considers the toxicity class of the pollutant. The tax levied may be 5, 50, and 500 times higher than the regular tax for classes 3 or 4, 2, and 1, respectively. The multipliers are arbitrarily chosen to stress the potential hazardousness of different pollutants.

The instruments used in Estonia are not analogous to those used in OECD countries, where noncompliance fees are fixed according to the profit the polluter receives if he does not comply with the ambient standards and exceeds the set emission limit.

In the case of disasters, explosions, or accidents, when hazardous emissions exceed the confirmed limits considerably, a noncompliance fee is levied on the polluter according to the toxicity class.

7.3 Practical Application of Emission Charges for Air Pollution Control

Regular confirmed emission charge levels are calculated on the basis of input resource characteristics, e.g., sulfur content of fuels, stack gas volume, the nature of the polluted area, and the content of up to 50 pollutants (Kallaste and Saare, 1991). Due to the lack of appropriate measuring equipment, emissions are calculated from the resource input data only; actual emissions are not measured. The emission charge is based on the following indicators:

- The volumes of the pollutants emitted.
- The size of the polluted area, which is assumed to be a circle with radius 80 times the stack height.
- The type of land use within the polluted area.
- The nature (hazardousness) of the pollutants emitted, compared with sulfur dioxide as a standard.

According to these indicators, the emission charges should perhaps be called ambient taxes.

Regular emission charges are imposed by municipal or county environmental protection boards on the basis of appropriate contracts. Polluters are supposed to pay the regular confirmed emission charges not later than the middle of every quarter. Payments in advance are possible. The polluter is informed beforehand of the amount of regular tax for the year.

Emission charges are applied for emissions into the air by stationary sources (stacks, municipal boilers, etc.) of pollution only. More than 300 polluters all over Estonia were obliged to pay emission taxes in 1991, more than 80 of which were in Tallinn. In fact, emission charges are levied on polluters whose emissions are higher than 200 tons per year or include very hazardous pollutants belonging to toxicity classes 1 or 2.

Due to the lack of an accounting system for the transport sector, emission charges for this sector have not yet been implemented, but some proposals to start earmarked fuel pricing are being discussed.

The present approach to the use of different charge rates is based on health risk, declaring that different pollutants present different risks to humans. All pollutants are classified into four toxicity classes. The first class includes pollutants that are most hazardous to humans; the fourth class includes modestly hazardous pollutants. This allows comparisons of the hazardousness of pollutants whose toxicities are quite different in the case of equal violations of the respective health standards (for more details, see Kallaste, 1989). The toxicity of one ton of SO_2 emitted is used as the basis for comparison; accordingly, emission charges per unit weight of emissions have been fixed (see *Table 7.1*).

The emission tax rate is calculated for each individual pollutant emitted. The total emission tax to be paid per year is the sum of individual taxes

Table 7.1. Information on the most relevant pollutants required for the taxation of emissions, January 1993.

Pollutant	Maximum permissible concentration MPC (mg/m^3) 24 hrs	Working area	Toxicity class	Emission tax rate p_i (EEK/ton)[a]	p_i for Tallinn ($/ton)
Sulfur dioxide (SO_2)	0.05	10.0	3	1.00	2.3
Nitrogen dioxide (NO_2)	0.04	2.0	2	2.50	5.8
Sulfuric acid (H_2SO_4)	0.1	1.0	2	2.24	5.2
Hydrogen sulfide (H_2S)	0.0008	10.0	2	7.91	18.3
Ammonia (NH_3)	0.04	20.0	4	0.79	1.8
Hydrogen fluoride (HF)	0.005	0.05	2	44.72	103.2
Formaldehyde (HCHO)	0.003	0.5	2	18.26	42.1
Oil-shale phenol	0.0007	0.5	3	37.80	87.2
Soot (C)	0.05	4.0	3	1.58	3.6
Cement dust	0.1	6.0	4	0.91	2.1
Particulates	0.15	6.0	3	0.75	1.7
Oil-shale flue ash	0.1	4.0	3	1.12	2.6
Petrol	1.5	100.0	4	0.06	0.1
Carbon monoxide (CO)	3.0	20.0	4	0.09	0.2
Phenol (C_6H_5OH)	0.003	0.3	2	23.57	54.4
Lead (Pb)	0.0003	0.01	1	408.25	942.7
Benzopyrene	0.000001	0.00015	1	57,735.03	1,332,347.0

[a]EEK = Estonian currency unit, the *crown*. 1 US$ = 13 EEK.

for each pollutant. The total tax for each pollution source is calculated according to the formula:

$$T = \sum_{i=1}^{z} P_i \times M_i \times K_a, \tag{7.1}$$

where T are annual tax payments for a single polluting enterprise, in crowns (the Estonian currency since June 1992); p_i is the emission tax rate for emitting one ton of a particular pollutant i into the ambient air, in crowns; M_i are actual emissions of pollutant i, in tons per year; z is the number of pollutants under study; K_a is a coefficient calculated on the basis of a spatial analysis of the areas of various land uses affected by the stack emissions.

The tax rate p_i is calculated as a root-mean value of two extreme standards – the working area and 24-hour standards for each pollutant (see formula 7.3) – based on the SO_2 tax rate (see *Table 7.1*). In the case of accidents, disasters, etc. the pollutant's toxicity class is taken into account.

The emission tax increases as multiples 5, 50, or 500, respectively, for pollutants belonging to toxicity classes 4/3, 2, and 1.

To calculate the coefficient K_a, different land uses are weighted according to the potential impacts of the pollutants. For example, open land in rural areas carries a weight of 1, industrial areas carry a weight of 20, and recreational areas, which are the most sensitive to pollution, carry a weight of 50. In the case of towns and other urban areas, the weight depends on the size of the affected area. This coefficient, reflecting the different areas affected by pollution, can be expressed by the formula:

$$K_a = S_{\text{rural}} + 20 \times S_{\text{ind.}} + 50 \times S_{\text{rec.}} + K_{\text{urban}} \times S_{\text{urban}}, \qquad (7.2)$$

where S_{rural} denotes the share of the area of water bodies, agricultural or forested land affected by air pollution; $S_{\text{ind.}}$ is the share of the industrial area; $S_{\text{rec.}}$ is the share of recreational areas; and S_{urban} is the share of territory of urban areas (towns, settlements, etc.) affected by air pollution.

The coefficient K_{urban}, in turn, depends on the population of a district as follows:

$K_{\text{urban}} = 20$, for less than 50,000 inhabitants;
$K_{\text{urban}} = 25$, for 50,000–200,000 inhabitants; and
$K_{\text{urban}} = 30$, for more than 200,000 inhabitants.

The area affected by the pollution source under study is calculated as a circle with radius $r = 80 \times H$, where H denotes the stack height.

The value of the coefficient K_a is fixed in the license granted to an enterprise on the basis of an agreed annual amount of emissions. The regular emission tax calculations are carried out each year and the changes in any parameters are recorded in a computer database.

To estimate the numerical value of the emission tax rate p_i for different pollutants the following approach is used. The coefficient p_i takes into consideration the differences in pollutants' daily and working area standards, and is calculated on the basis of the formula:

$$p_i = A / \sqrt{MPC_{i,\text{24hrs}} \times MPC_{i,\text{w.a.}}}, \qquad (7.3)$$

where $MPC_{i,\text{24hrs}}$ is the maximum permissible concentration of the pollutant i over a 24-hour period in the ambient air over settlements, in mg/m^3; and $MPC_{i,\text{w.a.}}$ is the maximum permissible concentration of the pollutant i in working areas, in mg/m^3. Its value is significantly higher than MPC_{24hrs}.

The numerical value of the coefficient A (we may call it simply a scaling factor) is chosen so as to make the SO_2 emission tax rate equal 1. Thus the SO_2 tax rate is used as the basis for comparison with the rest of the pollutants under study. For SO_2 $MPC_{24hrs} = 0.05$ mg/m^3 and $MPC_{w.a.} = 10.0$ mg/m^3. The value of the scaling coefficient A is set equal to $\sqrt{0.05 \times 10} = 0.707$. The values of the emission tax rates of the most common pollutants in Estonia calculated in this way are given in *Table 7.1*. For example, for HCHO (formaldehyde) the emission tax rate is:

$$p_{HCHO} = 0.707/\sqrt{0.003 \times 0.5} = 18.26.$$

This means that the tax levied on emissions of HCHO (which is significantly more toxic than SO_2; see *Table 7.1*) is 18 times higher than the SO_2 tax rate. In the case of Pb (lead):

$$p_{Pb} = 0.707/\sqrt{0.0003 \times 0.01} = 408.25.$$

In the case of pollutants with lower toxicity the appropriate emission tax rate is also lower. For example, for CO (carbon monoxide):

$$p_{CO} = 0.707/\sqrt{3.0 \times 20} = 0.09.$$

For these emission charge calculations, appropriate computer software has been developed and is used by every local county or municipal environmental protection board. At present, inventory data on emissions for the whole of Estonia by polluting enterprises are available in the computerized database of the Estonian Nature Management Information Center (1990). Pollution taxes have been calculated for emissions of about 50 pollutants in up to 500 enterprises throughout Estonia. More than 300 enterprises were actually charged in 1991.

7.4 First Results of the Emission Charges System in Tallinn

The total revenues from air pollution control for Tallinn city were about 970,000 roubles in 1991 (Tallinn Board of Environmental Protection, 1992). The expected revenues from regular emission charges were 950,000 roubles. The emission charge rate calculations were made for all major enterprises in Tallinn, but the obligations to pay were included in entrepreneurship contracts only in the case of 83 enterprises. The majority of revenues came from emission charges; only a small amount was obtained from other economic instruments applied, such as administrative charges, tax differentiation or noncompliance fees.

Table 7.2. Regular confirmed air pollution emission charges for Tallinn, 1991.

	Total emissions of pollutants (tons/yr)	Annual revenues (roubles)
Thermal power plant "Iru"	12,609.0	313,533.8
Heavy machinery plant	494.4	155,261.7
Municipal heating enterprise A	5,478.5	126,103.7
Enterprise "Estonian Phosphorite"	3,161.0	95,342.2
Baltic shipyard	525.5	34,024.1
Enterprise "Silikaat"	1,806.2	28,174.6
Tallinn metal works	429.9	26,118.9
Estonian fish treatment enterprise	148.8	20,700.8
Municipal heating enterprise B	148.8	15,063.0
Pharmaceutical plant	75.4	11,217.4
Total	25,651.9	825,540.2

Source: Tallinn Board of Environmental Protection, 1992.

The contribution of enforcement incentives to total revenues was also very small. This instrument obviously did not have the impact expected beforehand, for two obvious reasons. The polluting enterprises were very careful not to exceed the set limits, because they were aware of the new, very strict restrictions. Second, the output of many branches of industry has been permanently scaled down in recent years, so that far fewer people were working under the extreme conditions where peak emissions or accidents usually occur.

The majority of revenues consisted of contributions of the five to ten most polluting enterprises' emission charges only (see *Table 7.2*). The revenues for 1992 were very similar to those of 1991. In *Table 7.2*, three heat-generating enterprises appear in the list of ten major polluters. Their share in the total emission charge revenues to the Tallinn Fund is relatively large – about half of the total (Tallinn Board of Environmental Protection, 1992). The reason is that they use heavy fuel oil, which is rich in sulfur and no purification technology is used for desulfurization of stack gases, or for the removal of other pollutants. For further development of air pollution control, attention should therefore focus on the relatively small number of large or significant polluters.

The present approach takes into consideration the risk to human health, so that most attention is paid to highly toxic pollutants, whose effects are significantly more serious than those of other pollutants. Some polluters listed in *Table 7.2* discharge relatively small amounts, but probably of significantly more dangerous substances (e.g., pharmaceutical plants, heavy machinery plants), although their tax payments are still comparable with the rest (see *Table 7.2*). Highly toxic pollutants are liable to much higher charge rates (see *Table 7.1*). These are mostly emitted from outdated factories still using obsolete technologies. More technical innovation is needed to replace such manufacturing equipment.

Compared with water pollution emission charges, rates for air pollution control were greatly underestimated, and so they did not have the effect initially intended; this was also experienced in OECD countries in the past (OECD, 1991). Although regular emission charges are the most commonly used economic instrument, their application is greatly suboptimal, probably because they have not been fixed at a level high enough to achieve environmental goals. Hence, they provide significantly low level incentive effects and thus serve revenue-raising purposes only.

The high rate of inflation in the Estonian economy, which began in the early 1990s, has negatively affected the full-scale implementation of the new system of economic control for a wide area of environmental protection, including air pollution abatement. The inflation has created a critical situation where the incentive impact of economic control instruments, proposed at the rouble value of 1989/1990, turned out to be extremely low when applied in 1991/1992. It is important to note that the emission charges have not been raised in line with inflation, and the tax rates on pollutants have been set too low to be effective. The low tax level was a political decision made by the government to contribute to the survival of many enterprises, whose output was important for the republic.

Due to the high rate of inflation, the currency of that time (the rouble) rapidly lost its value and the biggest polluting enterprises were worried about the inflation rate and monetary reform. They were afraid that they would not be able to pay the regular confirmed emission charges during the coming year, so they paid them in full at the beginning of the fiscal year. Assessed from the point of view of the effectiveness of emission charges applied, this means, in fact, that the economic incentives in the form of emission charges did not work at all.

In January 1993, the Estonian government set new tax rates on the use of natural resources, water, and air pollution, as well as solid waste disposal (Government of Estonia, 1993). Following the monetary reforms of June 1992 when a new national currency unit, the Estonian crown (EEK), was introduced, the respective tax rates were revised (see *Table 7.1*). The exchange rate of 1 EEK = 10 roubles was fixed, but started to grow immediately. The numerical values of most of the new tax rates on resources and for pollution were fixed similar to those of rouble values. Comparing the 1992 and 1993 tax rates, one may interpret it as a tenfold increase in tax rates on most resources. At the same time, however, all prices, services, salaries, etc., also increased. As a practical result, it is difficult, if not impossible, to assess the effective increase in tax rates in general. During 1991/1992, the very high rate of inflation in the economy created a great deal of confusion in the accounting system.

7.5 The Estonian Environmental Fund

The Estonian Environmental Fund is currently dealing with the revenues from the application of economic instruments for pollution control. The Fund started in 1983, when the first step was made to introduce a strict and systematic taxation of environmental pollution. Estonia was the first among the 14 republics of the former Soviet Union to implement fines or noncompliance fees for polluting natural water bodies and air in settlements and cities, and to implement payments for the use of fresh water. The system was introduced in the 1960s and 1970s and enabled local environmental protection agencies to gain some experience in the field of environmental taxation.

The foundation of the Estonian Environmental Fund was accepted by the central government in Moscow as an All-Union economic experiment in the field of environmental protection and the rational use of natural resources. It was a real innovation in the environmental protection system of the former Soviet Union and was authorized to deal with environmental protection activities at the local level.

Before the Fund was established, noncompliance fees and fines were collected by local nature protection officers and channeled to the Ministry of Forest Management and Nature Conservation. About 90% of the revenues were centralized to the All-Union budget and only a minor share remained at the disposal of the local government. Administrative fines for the violation of nature protection regulations and nonrational use of resources were designed to generate revenues. In fact, fines for the pollution of water resources paid by polluting enterprises, agricultural farms, the army, etc., were the major sources of the Fund revenues. In recent years, before the reorganization of the Fund in 1990, the revenues totaled more than one million roubles per year; this was a quite remarkable sum for that time, although it represented only about 1% of the total expenditure on environmental protection in Estonia.

An unusual feature of this Fund was that more than 90% of the revenues remained in Estonia and were not taken away (were not centralized) by All-Union financial authorities in Moscow. It gave the Estonian environmental authorities for the first time a chance to use the money themselves. The Fund could be used to compensate for damage caused by pollution at the regional level, for investments in environmental technology, know-how transfer projects, scientific research, training, grants, international cooperation, etc.

In 1990 the Fund was reorganized into a nonbudgetary republican institution with the rights of a legal person. It was reoriented to handle the implementation of economic instruments for pollution control, and to collect revenues and use them as supplementary finance for environmental improvement. The Fund is attached to the Estonian Ministry of the Environment. Local funds are attached to county or municipal environmental protection boards. The activities of the Fund are run by a council, whose composition is fixed by the Estonian government, following proposals by the Ministry of the Environment.

The Fund is divided into: (1) the republican centralized fund, (2) 20 local (county or municipal) funds, and (3) special, so-called earmarked funds. At present, most of the Fund's revenues for air pollution control come from emission charges. The rest of the revenues, which consist mostly of fines and noncompliance fees, is to some extent unpredictable, since they depend on the polluters' own environmental strategies chosen each year under different economic conditions. The revenues from noncompliance fees and fines result from the sanctions applied in case of unplanned emissions into the air from leakages, accidents or disasters. "Chronic" or planned noncompliance is not common, because polluters often try to negotiate on the basis of regular confirmed emissions permits.

Revenues from emission charges and noncompliance fees are paid to the respective local county or municipal fund. The share of emission charges to the Fund is about 90%, and another six sources of revenue contribute about 10%. About half of the local fund revenues are given to the republican Fund and are used by both local and republican authorities to improve the air pollution situation at the national level. Every local environmental protection board, in cooperation with local authorities, has the right to draw up a specified list of the most hazardous air pollutants that present the most characteristic danger to local habitants or environment. The shares of water pollution, air pollution, and solid waste disposal tax revenues to the Fund have varied considerably during the first two years of implementation (see *Table 7.3*), resulting from fluctuations in the national economy. The share

Table 7.3. Revenues of the Estonian Environmental Fund, 1991 and 1992.

	1991		1992	
Source of revenue	Roubles (1,000)	%	EEK (1,000)	%
Water pollution emission charge	34,291	59.2	3,756	50.2
Air pollution emission charge	8,432	14.6	589	7.9
Waste disposal charge	13,491	23.3	1,135	15.2
Other sources	1,651	2.9	2,003	26.8
Total	57,865	100.0	7,483	100.0

Source: National Board of Waters and the Environment, Environmental Data Centre, 1992, 1993.

of air pollution emission charges has decreased significantly, mostly due to the rapid decrease in oil shale-based energy production, which is based on outdated technology without sulfur purification.

As for the centralized republican Fund, its revenues are generally based on a redistribution of local fund revenues. The use of the republican Fund's means is fixed by the board of the Fund. In principle, local funds have the right to make their own decisions on the uses of the revenues, which remain at their disposal after distribution for environmental improvement.

7.6 Comments on the Implementation of Economic Instruments

A number of aspects should be mentioned in connection with the establishment of an empirical framework for economic instruments in air pollution control in Estonia. For a long time there has been no significant economic incentive for pollution abatement measures, nor investment in environmentally benign technologies or flue gas cleaning equipment. Direct administrative regulations, in combination with some first elements of economic instruments for more efficient pollution control, have been relatively ineffective until recently.

In general, estimates of environmental improvement costs have been made neither on the basis of the value estimates of damage caused by emissions of pollutants, nor on the full cost of compensation for damage to particular components of the environment. They are not based on market economy principles nor deduced from analyses of their cost-effectiveness.

The total desired annual revenue of the Fund that could ideally be used for pollution control, i.e., guaranteeing investments and labor for pollution abatement activities (according to the logic of the centrally planned economy) has simply been fixed. For example, experts estimated that the total revenues needed were more than 6 million roubles at 1989/1990 prices. This estimate was used to calculate the charge rates per unit of emissions. Since enormous changes have taken place in all sectors of the national economy in recent years, it is obvious that a new, revised method of calculating the tax rate basis needs to be developed as soon as possible. Since the taxes were introduced, only a few revisions of tax rates have been made, most recently in January 1994. This means that the principles of the planned economy are still being used.

The revenues available to the local funds vary significantly by county or municipality, i.e., regionally. The funds of four counties and municipalities, out of 20, contributed about 82% of total revenues from air, water, and solid waste taxes. These four regions are Tallinn city and three counties or towns in the northeast of Estonia where air pollution charges are also high. It is quite normal that these local funds are not able to use all the revenues they receive in the most efficient way, which is one reason why they are redistributed among local funds.

7.7 Conclusions

Many changes have been made in the economic structure of Estonia in order to achieve the more efficient use of natural resources and make considerable reductions in pollution loads. Recently a new system of policy instruments for more efficient environmental control was introduced. The implementation of new economic instruments for air pollution control, such as emission charges since January 1991, has been a major step forward. The newly established tax system has various aims, the most important of which are:

- To reduce heavy pollution loads in the most critical areas, and to reduce specific forms of environmental damage using different types of market forces.
- By implementing the "polluter-pays" principle, to select the most efficient and low-waste production technologies, thus stimulating polluting enterprises to reduce their emissions and at the same time to conserve raw materials.
- On the basis of resource pricing, to change the free-input approach that has provided the preconditions for the inefficient use of resources and has given rise to unacceptable levels of air pollution.

- To accumulate earmarked money for environmental improvement activities.

In contrast with the practices in neighboring Nordic countries, such as Finland, Sweden, and Norway, in Estonia the revenues from pollution taxes are used exclusively to improve the state of the environment.

The introduction of a system of market instruments has been a kind of economic experiment, since until recently, Estonia was ruled by command economy principles, where resources had no price before they were produced. The free use of natural resources caused inefficient production levels with overuse of resources, and resulted in a high output of wastes. Flue gas cleaning technologies have not been widely used until now, since there have been no real economic incentives to do so.

There is no time left to postpone the introduction of the economic instruments and resource pricing. Further deterioration of the environment should be prevented using every possible means. However, since the whole economic system itself is in transition, the use of economic instruments cannot produce significant effects. Estonia has started to take measures to reduce the further deterioration of the environment, and this should serve as the basis for sustainable development. For quick results in environmental improvement, it is obvious that Estonia needs subsidies and long-term loans from World and European institutions. A carefully designed master plan for environmental policy should be developed for Estonia in the near future.

The new system of economic control of environmental protection is a "mixed approach", combining elements of economic instruments, resource pricing and direct regulations. It is also mixed in the sense that "old" approaches and regulations are still in force. The levels of charge rates are based mostly on the empirical knowledge of experts and principles from the Socialist period. Classical economic theory and fundamental calculations have not been used to set the charge levels. This, in turn, may be one reason for the relatively large differences (up to tenfold) in the revenues of various spheres of implementation of economic instruments – in air pollution, water management, and waste disposal control.

The main conclusion of this chapter is that the pollution charges in Estonia have been set too low to act as a real economic incentive for polluters. In air pollution, charging polluters has had little impact on their economic behavior. The revenue-raising character of the charges is still dominant. For instance, the largest contributions to the Estonian Environmental Fund were made by the two oil shale-based power plants in northeastern Estonia; in 1992 they paid up to 18% of total pollution charges. However, the air pollution tax formed only a small part (0.08%) of the price of electricity (National Board of Waters and the Environment, 1993).

The extremely complex transition from a centrally planned economy to a market-oriented system in Estonia has reduced the expected environmental effectiveness of the market instruments implemented. The very high inflation rate and restructuring of most sectors of the economy have significantly reduced the expected revenues from the new system of economic control of pollution. The cost-efficiency of economic incentives is improving, but has not yet reached its optimal level. Still, the pollution charges stimulate the cost-efficient allocation of pollution abatement even at their present suboptimal levels. When the national economy and accounting system broke down in the early 1990s and rapid inflation started, the traditional principles of assessing the efficiency of economic instruments were no longer applicable. An emotional style is used at present when assessing the effectiveness of pollution charges applied. The data on environmental expenditures are still incomplete and often incompatible. The first attempts to build up a new national accounting system were made only recently.

The very low pollution tax levels resulted from political decisions made by the government to assist many enterprises to adapt to the changed economic situation. This has created a situation where the revenues from economic instruments and resource pricing (designed at the rouble value level of 1989/1990) were relatively low from the start. The government is increasing the tax rates stepwise; the most recent revision was in early 1994. As the level of inflation remains high and pollution taxes have still not been adjusted, optimal marginal costs are lower, and less pollution is controlled.

An important result expected from the implementation of economic instruments for air pollution control is a gradual shift in production and consumption into a more efficient and environmentally friendly direction, which will form the basis for sustainable development in Estonia. The transition to a market economy is still in progress. Some elements of the centrally planned economy continue to exist, preventing the cost-effective operation of economic incentives. For example, soft budget constraints, fixed prices, centrally granted investments, etc., are used in some important sectors of the national economy. Administrative regulations and economic incentives have a long way to go before they achieve cost-effectiveness criteria. It is obvious that to establish the principles of resource pricing and economic control of pollution similar to those used in other European countries, considerable effort is still needed.

Acknowledgments

The author very much appreciates the useful comments by Professor Thomas Sterner from the University of Gothenburg, and Professor Tomasz Zylicz from the University of Warsaw, and two anonymous referees. Special thanks to Ger Klaassen from IIASA whose editorial work improved the manuscript significantly.

References

Bye, B., Bye, T., Lorentsen, L., 1990, Taxation of Fossil Fuels: The Impacts on Economic Activity, Energy Markets and Pollution, in K. Halsnaes and G.A. Mackenzie, eds., *Nordic Workshop on Integrated Energy and Environmental Modelling*, Risø National Laboratory, 15–16 February 1990, *Nord* **74**, Nordic Council of Ministers, Roskilde, Finland.

Estonian Nature Management Scientific Information Center, 1990, *Environment '89*, Tallinn, Estonia.

Government of the Republic of Estonia, 1990, Regulation No. 237 of 20 November 1990, *On the Regulations Concerning the System of Fines and Payments for Nature Management*, RT, Tallinn, Estonia (in Estonian).

Government of the Republic of Estonia , 1992, Regulation No. 7 of 9 January 1992, *On Changing Regulation No. 237 of the Government of Estonia of 20 November 1990*, RT, Tallinn, Estonia (in Estonian).

Government of the Republic of Estonia, 1993, Regulation No. 5 of 8 January 1993, *On Changing Regulation No. 7 of the Government of Estonia of 9 January 1992*, RT, Tallinn, Estonia (in Estonian).

Kallaste, T., 1989, How to Assess the Level of Pollution, *Fennia* **167**(1):73–85.

Kallaste, T., 1991, *The Environmental Problems in Estonia and Some Ways to Solve Them*, Working Paper, No. 76, Research Institute of Northern Finland, University of Oulu, Oulu, Finland.

Kallaste, T., 1993, *Taxation and Environment: Case Study on Estonia*, Paper presented on the OECD Workshop on Taxation and Environment in European Economies in Transition, OECD, Paris, France.

Kallaste, T., and Saare, L., 1991, Control and Assessment of the Environmental Pollution Situation in Estonia, in J. Fenger, K. Halsnaes, H. Larsen, H. Schroll, and V. Vidal, eds., *Environment, Energy and Natural Resource Management in the Baltic Region*, 3rd International Conference on System Analysis, May 7–10, Copenhagen, *Nord* **48**, Nordic Council of Ministers, Copenhagen, Denmark.

Kallaste, T., Roots, O., Saar, J., Saare, L., 1992, *Air Pollution in Estonia 1985–1990: Environmental Report 3*, National Board of Waters and the Environment, Environment Data Center, Helsinki, Finland.

Ministry of the Environment, Republic of Estonia, 1991, *Environment '90: A Survey*, Tallinn, Estonia.

National Board of Waters and the Environment, Environmental Data Center, 1992, *Estonian Environment 1991*, Environmental Report 4, Helsinki, Finland.

National Board of Waters and the Environment, Environmental Data Center, 1993, *Estonian Environment 1992*, Environmental Report 6, Helsinki, Finland.

Nicolaisen, J., and Hoeller, P., 1990, *Economics and the Environment: A Survey of Issues and Policy Options*, Working Paper No. 82, Department of Economics and Statistics, OECD, Paris, France.

OECD, 1992, *Environment and Economics: A Survey of OECD Work*, Organisation for Economic Co-operation and Development, Paris, France.

OECD, 1991, *Environmental Policy: How to Apply Economic Instruments*, Organisation for Economic Co-operation and Development, Paris, France.

Opschoor, J.B., and Vos, H.B., 1989, *Economic Instruments for Environmental Protection*, OECD, Paris, France.

Tallinn Board of Environmental Protection, 1992, *Yearbook 1991*, Tallinn, Estonia (in Estonian).

Chapter 8

Application of a Pollution Fee System in Russia

Alexander Golub
Market Problem Institute, Russian Academy of Sciences
Moscow, Russia
Elena Strukova
Commission for Natural Resources
Moscow, Russia

Abstract

This chapter analyzes the development of a new management system for environmental protection in Russia during the period of transition to a market economy. Under the former command-and-control (CAC) system the share of abatement investments increased, and emissions actually declined, but environmental problems were not resolved, and environmental quality standards were not achieved. The CAC system collapsed, together with the centralized system of national economic management, because it was based on budgetary resources. It was replaced by economic mechanisms, including a system of pollution fees. This chapter traces the development of the pollution fee system, and explains the reasons for its inefficiency. Although actual emissions have declined, the level of pollution per unit GNP has increased, and the share of environmental investments in GNP has fallen. The major reasons for the weaknesses of the system have been severe cuts in state funds, the low level of the fees charged, the inadequate fee collection system, and the overall lack of markets in Russia. However, the institutional foundations for an environmental management system have been laid, and the system is likely to become more effective as the market economy develops.

Key words: pollution fees, damage estimation, command-and-control system, environmental fund, environmental protection investments.

8.1 Introduction

This chapter investigates the system of environmental management in Russia, particularly economic instruments. The way the pollution fee system was formed and is operated is analyzed. The analysis covers 2–4 years starting from an experiment conducted in 1989–1990 in a limited number of regions, up to the current results of the operation of the new pollution fee system throughout the country. Focusing on an analysis of air pollution management, we examine the ways in which enterprises are being encouraged to accept responsibility for environmental protection, and the incentives provided by economic instruments for the reduction of environmental pollution.

The administrative system of environmental management that existed until 1990 reflected the overall system of Soviet economic regulation. However, the recent economic reforms put an end to the administrative mechanism of regulation in its old form, and a new regulation system corresponding to general market relations was introduced, based on economic instruments. The main issues we address here are whether the new system has created incentives for polluting enterprises, what have been the positive results of the introduction of the system; and what needs to be done to increase its efficiency. Of course, the short period of operation of this new system is insufficient to allow a deep and full analysis. We therefore analyze indirect factors, compensating for the lack of empirical material and data by expert estimates.

Such an investigation is relevant for the creation of market systems in Russia. Economic mechanisms with an appropriate system of environmental standards will soon be the only tools for ensuring the environmental responsibility of enterprises. The privatization of public enterprises has given many rights to the new owners, but the weaknesses of existing economic and legal systems could create a disproportion between the rights and liabilities of the owners, and could have negative consequences for the environment. In this chapter we examine whether the new system of pollution fees provides sufficient incentives for firms to invest in environmental protection. What could make them comply with emission standards other than a system of pollution fees? In answering these questions we take into account the current market situation in Russia.

The command-and-control system did have some positive results for environmental protection, however. It is therefore interesting to compare the effectiveness of the old and new systems, although under the current deep economic crisis it is very difficult to judge the merits of the new system. We first need to distinguish different tendencies.

8.2 The Command-and-Control System of Environmental Management

The command-and-control (CAC) system involves a number of elements, including environmental quality standards; planning and financing of environmental activities; and monitoring and controlling environmental quality. The main features of this system were determined in a set of legislative acts adopted by the USSR Supreme Soviet: the Basic Land Management Act (1968), the Water Management Act (1970), the Mining Management Act (1975), and the Forests Management Act (1977). In 1980 acts about the Protection of the Atmosphere and the Protection and Utilization of Wildlife were also adopted. These acts formulated general rules for the exploitation of the environment and natural resources. Environmental protection goals were set in the Public Health Act (1969), and the necessity for environmental protection was first mentioned in the constitution of the USSR (the Main Law, 1977). In all of these acts the responsibilities of the various ministries and organizations in the field of environmental protection were specified.

8.2.1 The system of environmental quality standards

In accordance with the Public Health Act (1969) the Ministry of Public Health was obliged to work out and adopt environmental quality standards. Current ambient standards (USSR Council of Ministers, 1980) played a key role in the system of standards in the USSR, which was based on maximum permissible concentrations (MPCs) of hazardous substances in the environment (atmosphere, reservoirs, soils).

It was considered that the MPCs, if adhered to, would preserve and maintain the properties of the environment, i.e., that the levels set in the MPCs were sufficient to protect human health, plants, and animal life. Announced in 1969 and introduced in 1978 (USSR Council of Ministers, 1978), the MPC standards corresponded to medical requirements and were very severe (often more stringent than in the EC or the USA). For SO_2, for example, the MPC was 0.05 mg/m^3 in the USSR, 0.3 mg/m^3 in Switzerland, and 0.26 mg/m^3 in the USA. Examples of MPCs for towns are shown in *Table 8.1*. Such standards were set for more than 200 substances on the

Table 8.1. Examples of maximum permissible concentration (MPC) standards for urban areas (mg/m^3).

	Annual maximum	Daily average
SO_2	0.50	0.05
Cl	0.10	0.03
Hydrogen sulfide	0.008	0.003
CO	6.01	2.01
NO_x	0.3	0.1
Phenol	0.30	0.01
Pb	–	0.0007
Hg	–	0.0003
Arsenic	0.30	0.01
Dust (nontoxic)	0.50	0.15

basis of medical criteria only, without regard to economic and other factors, which explains their severity.

Such rigid standards were practically unattainable, however. According to data for the early 1980s, average MPC standards were exceeded by factors of 2.4–2.5, particularly in urban areas.[1] On the basis of compliance with the MPC standards maximum permissible levels of emissions (MPLs) for enterprises (stationary sources of pollution) were fixed. (For mobile sources, concentrations of exhaust gas emissions were also fixed, although in practice, only the CO emissions of vehicles were controlled). These standards were to be followed by both new and existing enterprises. They were calculated by the State Hydro-Meteorological Committee and the Ministry of Public Health and introduced from 1 January 1980, in accordance with the Atmospheric Protection Act. In practice, enterprises' actual emissions were often substantially higher than the MPLs.

For economic and technical reasons, and because specific local conditions were not taken into account, it was unrealistic to hope for compliance with these standards; indeed, the majority of industrial enterprises would have had to close down if the MPL requirements had been accurately observed. Thus temporary coordinated standards (TCS) were introduced; these were higher than the MPLs but lower than actual emissions. They were set for existing enterprises according to individual circumstances (new enterprises were required to comply with the MPLs). The possibilities for enterprises to reduce their emissions step-by-step to comply with the MPLs were taken into account in these standards. Special emission reduction programs were elaborated for large enterprises, taking into account the costs of all abatement measures. The temporary standards were thus a compromise between the environmental and economic goals of Soviet society.

MPLs and TCS were fixed only for the largest enterprises at first. Several years after their introduction (before the pollution fee system experiment) only 14%–18% of all enterprises had complied with the standards.

8.2.2 The system of planning and financing of environmental activities

From 1982 environmental planning became part of production plans of enterprises, general plans for urban development, and regional schemes for the location and development of industrial complexes. On the basis of the plans, enterprises were to achieve, step-by-step, compliance with the MPLs. First, enterprises had to attain the TCS and then the MPLs. This process had to be completed by the year 2005, when the goals of the Long-Term Environment Program (LTEP) were to be realized (USSR State Committee on Scientific and Technological Development, 1980); i.e., compliance with MPC standards throughout the USSR on the basis of the MPLs for emissions by all enterprises. Thus for all such enterprises special investment plans were drawn up; the investments had to be sufficient first to reach TCS and then MPL standards. These investment plans formed the main basis for abatement activities of enterprises. They were worked out by the responsible ministries of the USSR and Soviet Republics on the basis of the tasks of regional authorities (executive authorities of regional and local Soviets), as well as the USSR and Republican plans for the protection of water, air, land, and forests, as well as national parks, wildlife, fish stocks, the rational use of resources, and for the introduction of pollution abatement equipment.

Between 1981 and 1990 abatement activities were outlined in five- and one-year plans for the socioeconomic development of enterprises. State investments for environmental protection were fixed by the USSR State Planning Committee. The Long-Term Environment Program (LTEP) was to provide the general framework document, and at the regional level Regional Complex Schemes for Environmental Protection were drawn up for the most polluted and unique regions (e.g., Lakes Ladozhskoe and Baikal). From 1984 onward the required investments were also financed from the state budget after a special decision of the government. At the level of the enterprise, abatement expenditures also were stipulated in the plans. This money was given to enterprises from the budgets of the corresponding industry ministries, in consultation with the state organizations responsible for environmental protection. For infrastructural projects (such as sewerage) funds were provided from local budgets. Thus, at the highest level the total abatement expenditures were determined by the USSR State Planning Committee. Part of this money was then distributed among ministries and later among enterprises according to their plans, and another portion of the money was distributed among regions. The flows of funds could intersect; for example, enterprises could obtain funds for abatement from both the regional environmental programs and from ministries.

The final result of this resource allocation should have been the achievement of ecological standards. Every year enterprises received, from the upper level, emissions reduction plans with corresponding investments and limits on investment resource utilization (capacities of building firms, materials, etc.), which were the most important. Thus the financial and physical aspects of environmental activity were well planned.

8.2.3 The environmental quality monitoring and control system

Monitoring and control functions were performed by a number of state organizations. The State Hydro-Meteorological Committee was responsible for the atmosphere and a special Air Protection Inspectorate was created. Within the Ministry of Public Health there were special hygienic services which, before the emergence of the State Environmental Committee in 1989, were responsible for maintaining air quality. In particular, they had to stipulate MPLs and TCS for enterprises and to supervise compliance with them; grant emission permits; and organize environmental monitoring systems for the collection of data on emissions. These organizations included inspection departments, but they were unable to influence the polluters. Their main tools were to impose penalties (up to a maximum of 100 roubles), to institute criminal proceedings (very rare), or to close down enterprises (almost never).

8.2.4 Reasons for the collapse of the CAC system

Many Russian specialists have commented (Gofman, 1977; Gofman and Gusev, 1977) that the CAC system did not create enough incentives for the improvement of abatement activity, but nevertheless there was a steady reduction in harmful emissions per unit of production in the former USSR after 1980. This is apparent from *Figure 8.1.* If the CAC system had continued its operation, then with regard to the Long-Term Environment Program

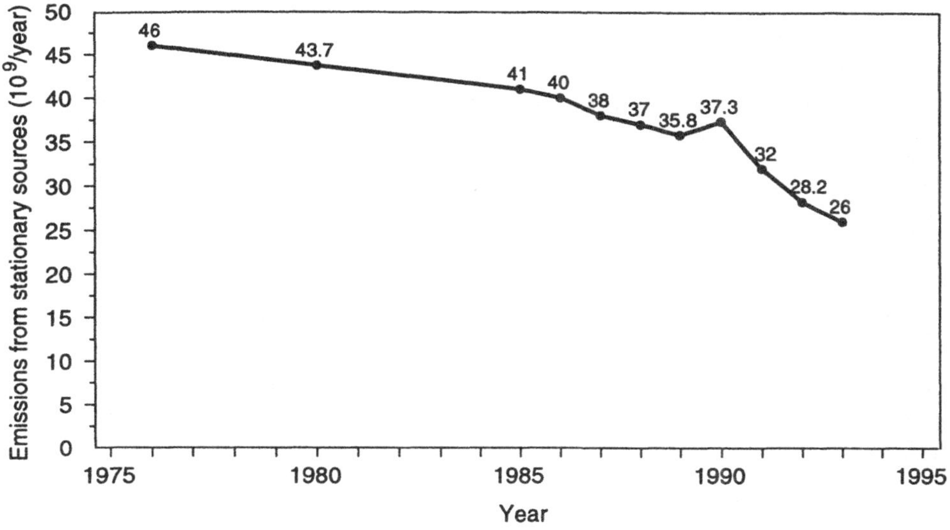

Figure 8.1. Dynamics of air pollution from stationary sources. Source: Calculated according to USSR Statistical Committee, 1985, 1988, 1989, 1990.

Table 8.2. Air emissions planned by LTP (million tonnes).

	1980	1985	1990	1995	2000	2005
Total	110.8	105	94	85.4	70.7	53.2
of which stationary sources	72.8	68.3	57.3	50.8	42.1	32.6

Source: Russian Council of Ministers, 1991.

(USSR State Committee on Scientific and Technological Development, 1980) MPC standards would have been achieved by 2005, although much higher investments would have been needed. According to the estimates of the experts who designed the program, real abatement investments would have needed to be two or three times higher. Although the CAC system did achieve some positive results, it was rather expensive. The environmental planning and management systems were not oriented toward achieving the MPC standards at minimum cost. *Table 8.2* illustrates the probable decrease in emissions over Soviet territory if the Long-Term Environment Program had been realized.

The CAC system required environmental quality standards to be reached with the help of state abatement investments. But with the exhaustion of the state budget resources the main means for meeting the standards were no longer available. The CAC system created no incentives for purification or abatement activity. It was unable to survive without state financial resources, leading to the waste of budget resources.

The CAC system collapsed in 1991–1992, together with the general system of planned economic management. The CAC in environmental management had been created in 1980–1985 and it disintegrated together with the rest of the planned economy.

The CAC did not solve any environmental problems of the former USSR, but nevertheless prevented, to some extent, further deterioration in environmental quality. Although its influence on enterprises was not significant, it was tangible. In the following, we examine the consequences of the disintegration demise of the CAC system, and whether the new system will be more effective in improving the pollution dynamics after the economic crisis in Russia.

8.3 The Pollution Fee System

In 1988, All-Union and Russian Ecological Committees were set up with responsibility for the management of pollution control. In 1989 the All-Union Committee organized an experiment to introduce pollution charges, and in 1991 the system was extended throughout the territory of the former USSR. In 1993 the Russian Ecological Committee was transformed into the Ministry of Ecology and Natural Resources, with a wider range of functions. In 1992, the Russian parliament passed an Environmental Protection Act and the Ministry continued its efforts to create an adequate environmental management system for the new socioeconomic and political conditions.

At the end of 1980s it became clear that the command-and-control system could not survive the transition to a market economy. The need for a new system of environmental protection and management also became obvious. A pollution fee system was therefore introduced, with two levels of fees: a basic rate applied when enterprises operated within their permitted emission limits, and an increased rate (five times higher) applied for emissions above these limits. Regional authorities can adjust the rate with correcting coefficients to take into account specific local circumstances.

For each enterprise, the environmental authorities define permitted volumes of emissions, usually temporary standards, that were fixed in the license. Then a legally binding agreement for natural resource utilization is drawn up reflecting the order of pollution fees to be paid. Enterprises are committed to pay the basic rate for emissions within the permitted level, and the increased rate for emissions above it.

Together with the introduction of the pollution fee system, environmental funds were also established at three levels: local, republic (regional), and federal. Pollution fees contribute to these funds. According to data from the Ministry of Environmental Protection and Natural Resources, in 1991 a total of almost 1 billion roubles was collected in the form of pollution fees, penalties, and fines (imposed together with fees). The main source were fees for pollution within the permitted emission limits (65%). Fees for pollution above the permitted levels contributed 14.8%, fines 9.3%, penalties 2.6%, and other sources 8.3%. Penalties and fines constituted only 12% of the total. These data demonstrate that fees for pollution are the main contributors to the environmental funds.

8.3.1 The experiment to introduce pollution fees in 1989–1990

The CAC systems in the US market economy and in the USSR operated in very different ways. The main efforts of the US government have been to specify and control compliance with ambient and technology-based standards together with the necessary legal enforcement. The main task of the Soviet government, on the other hand, was to ensure the proper distribution of abatement investments. Standards were fixed but almost never met in the USSR. The basis of the Soviet CAC system was the centralized provision of materials and financial means for environmental investments, and this system collapsed together with the rest of the system of centralized planning and finance.

The administrative mechanisms of pollution control were inefficient, but scientists and managers understood that it was likely to become much more inefficient in future during the transition to a market economy. New methods of environmental management were badly needed. The USSR State Ecological Committee decided to put into operation a pollution fee system and in 1989–1990 they conducted a corresponding experiment, involving about 50 regions, including urban areas such as Moscow, Dnepropetrovsk, Zaporozhje, Nizhniy Tagil, and other towns and regions with poor environmental records.

To give an idea of the scale of the experiment, it was conducted in towns such as Moscow and St. Petersburg, where the main part of industrial production is concentrated, and where 10% of the population lives. The annual levels of emissions into the air were 1.15 million and 608,300 tonnes, respectively. Other regions involved in the experiment included Tver, Vologda, Cherepovets, Kostroma, and Jaroslavl in the European part of Russia. These are industrial regions with serious levels of air pollution: Cherepovets in the Vologda region (741,700 tonnes per year) and Jaroslavl (309,700 tonnes per year). The biggest metallurgical plant in Russia is located in Cherepovets.

The geographical diversity of the regions involved in the experiment was reinforced by the inclusion of Siberian regions such as Kemerovo (202,000 tonnes of emissions into the air) and the Ural town of Nizhniy Tagil (668,500 tonnes of emissions into the air). Regions with unique natural features were also included; e.g., Lake Baikal in the Irkutsk region (156,600 tonnes of emissions into the air). The wide diversity of participating regions with different geographical, climatic, natural, economic conditions, and industries ensured that the experimental results would be truly representative.

In the experiment two methods for calculating and collecting pollution fees were tested. The first was based on measuring economic damage, and the second on calculations of the regional expenditure required for pollution control.

This first attempt to introduce pollution fees (for air and water pollution, and solid waste discharges) was based on a decision of the USSR Council of Ministers. The question of which regions would participate in the experiment was resolved by local authorities, who had to request to take part in the experiment. However, the imposition of effluent fees had no legal basis; every enterprise could therefore refuse to pay, and they very frequently did so. The environmental authorities then had to find indirect ways to persuade them to participate; for instance, if the enterprise had to submit a project for the authority's approval, it would not receive an answer until all pollution fees were paid.

We now examine how the fees were calculated. The first method was based on damage estimation. To calculate damage a simplified reduced method to standardize the procedure was used. The most common approach was to calculate the damage on the basis of a single formula:

$$u = v * G * M, \tag{8.1}$$

where u is the estimate of damage; v is the monetary assessment of the damage caused by 1 tonne of conventional aggregated emissions; G is a special coefficient, to take into account particular features of the region; and M is the total volume of aggregated emissions:

$$M = A_1 * m_1 + A_2 * m_2 + \dots + A_i * m_i, \tag{8.2}$$

where m_i is the total volume of emissions of substance i; A_i is the coefficient of the relative danger of substance i.

By means of the coefficient A it was possible to aggregate various harmful substances and discount them to conventional types of emissions. Special rules were elaborated to help regional authorities calculate rates of fees (see Gofman *et al.*, 1986).

The collection of fees calculated on the basis of this method presented a number of problems. In none of the cities in which this method of fee calculation was implemented, was the actual charge correlated with the calculated value of damage. Enterprises were simply unable to pay the estimated sums. Thus a second variant for calculating the fees was proposed. The expenditures necessary for the realization of the regional pollution control program (K) were defined by the regional Environmental Protection Committees. Then the payment for each polluter was calculated as a share of the total abatement costs using the formula

$$P = (M_j/M)K, \tag{8.3}$$

where P is the emission fee; M_j are the emissions of polluter j; M is total volume of emissions; and M_j/M is the share of enterprise j in the total volume of emissions.

The payments calculated by formula (8.3) were 10–15 times lower than the level of incurred damage, calculated by formula (8.1).

To create the necessary incentives for enterprises to curtail emissions a special system of limits was proposed. For emissions within the MPL standards, payments would be calculated using formula (8.3): the common tariff. For emissions above the standards, the penalty tariff applied.

The main idea behind calculating the fees on the basis of formula (8.3) was that for each territory the expenditures needed to realize the environmental programs were known, and were different for each region. To calculate penalty fees one needs to know the level of abatement expenditures of each enterprise. Unfortunately, in most cases it was impossible to obtain such data, so that in 1991 the method was simplified. By a decision of the Russian Council of Ministers (1991), single rates of payment were adopted for the whole of Russia (*Tables 8.3* and *8.4*). This decision was extended for 1992, but according to local policy the rates could be changed. Local authorities could use different approaches for different polluters, especially taking into account regional conditions.

The experiment demonstrated the possibility of collecting pollution fees, but also revealed many problems. The main problem was the absence of a legal basis for collecting fees, since the decision of executive authorities was

Table 8.3. Fees for emissions of some air pollutants (roubles/tonne).

Pollutant	Within permissible limits	Above limits
NO_x	55.01	263.39
SO_2	66.00	316.00
Pb	10,999.89	52,666.14
Dust from timber, lime, and coal	22.01	105.39
CO	1.09	5.21
Benzopyrene	3,300,000.00	15,800,000.00

Source: Russian Council of Ministers, 1991.

Table 8.4. Fees for discharges of some water pollutants (roubles/tonne).

Pollutant	Within permissible limits	Above limits
Aluminium hydrate	8,870.00	46,934.00
Aniline	4,435,000.00	23,467,000.00
Vanadium	443,500.00	2,346,700.00
Oil	44,350.00	234,670.00
Arsenic	8,870.00	46,934.00
Phosphates	17,740.00	93,868.00

Source: Russian Council of Ministers, 1991.

only an indirect base. A polluter who refused to pay could not be forced to do so. This problem was resolved in 1991, however, when the pollution fee system was introduced throughout Russia.

8.3.2 Russian system of pollution fees in 1991

According to the results of the experiment it was decided that pollution fees would be imposed throughout Russia from 1 January 1991 onward. The fees would apply to air pollution from stationary sources; air pollution from mobile sources (vehicles); pollution of water; and the disposal of solid wastes.

The legal basis for the pollution fee system was established by a decision of the Russian Council of Ministers: "The polluting emission fee system and its implementation" (Russian Council of Ministers, 1991). In this decision, specific fees for 1991 were adopted, and it was stressed that they would apply to all polluters, regardless of their institutional positions. It was shown that fees could be reduced at the regional level in light of specific environmental and economic conditions. A special provision was created to include expenditures for environmental protection for the amount of fees necessary.

Supplements to this decision fixed two types of fee rates for water and air pollution (*Tables 8.3* and *8.4*): for technologically approved discharges (emissions within the standards); and for emissions above the standards, which were five times higher than those for emissions within the standards.

Pollution fees were fixed for emissions of 211 substances into the atmosphere, and for discharges of 92 pollutants into water bodies. Moreover, fees for exceeding the standard emissions were fixed for 33 pesticides, although fees were not levied on discharges of pesticides within the standards because of the unstable economic situation in the agricultural sector.

An essential feature of the pollution fee system was its uniform character, independent of the region or the location of the source of the pollution. Obviously, the regions had different abilities to absorb and compensate for pollution, so that special correction factors were introduced to take into account specific regional conditions. Different levels of regional aggregation for air and water pollution were chosen: only 11 regions were distinguished for air pollution (the Far East region had a correction factor of 1; the Ural region had a correction factor of 2), whereas for water pollution the division was more detailed, with 99 water regions.

The correction factors resolved the problem only partially, however. As mentioned above, one of the main ideas behind the pollution fee system was to create independent sources of finance for regional environmental programs. Thus the fee rates should have been linked to the regional demand for financial resources for such programs.

8.3.3 The current pollution fee system in Russia

The current system of pollution fees was formulated after the adoption of the 1992 Environmental Protection Act (Russian Supreme Soviet, 1992). Chapter III of article 20 of the Act provided the legal basis for the imposition of fees for pollution. As before, rates of payment were set by the government, but could be changed by the executive authorities. In August 1992 the government adopted a decision (Russian Council of Ministers, 1992) on a procedure for calculating fees for pollution and other harmful influences, and the fee levels.

The possibilities for regional differentiation of fees were increased. Regional environmental authorities, in consultation with the federal government, could define their own pollution fees. A new system of payments with regional differentiations was put into operation in 1993. At the same time, fees were introduced for other types of harmful influences, such as noise, vibration, electromagnetic fields, radiation, etc.

Indexation of pollution fees was undertaken five times in 1992, and twice in 1993 (in 1993 above-standard emission fees were increased 25 times whereas within-standard fees were increased less often). In 1993, the total fee payments of both above- and within-standard emission fees were ten times higher than in 1992. In 1994 it is planned to increase pollution fees by a factor of 15. Thus total pollution fees will have been increased by 150 times over three years, even though the rate of inflation is about 300% per year (the real dollar:rouble exchange rate was 50:1 in 1992, approximately 500:1 in early 1993, and 1300:1 at the end of 1993, in early 1994 it was 1600:1). Thus the total impact of the pollution fees has decreased dramatically.

Of course, the growth of payments has not been equal to the rate of inflation, and this has been the main reason for the sharp reduction in the share of environmental protection expenditures in the Russian GNP.

Another type of economic instrument is taxation policy. The Taxation Act (Russian Supreme Soviet, 1991) stipulated that 30% of profits allocated for environmental improvement purposes would be exempt from profit tax. In the Environmental Protection Act even wider exemptions were also outlined. Experience has shown, however, that these exemptions could be used only together with corresponding changes in other laws. The Russian Finance Ministry opposes these exemptions, because they will reduce tax revenues; the budget is currently showing a deficit.

It would be interesting to examine the incentives arising from existing tax exemptions, but unfortunately little information is available. However, experts' estimates show that under the present system of taxation and fees, these exemptions do not create enough incentives for the allocation of resources to the environment sector. Even if the tax exemptions were to cover 100% of environmental investments, and if enterprises were also exempt from value added tax, then the fees would cover only 48% of the necessary abatement expenditures.

8.4 Estimation of the Efficiency of the Pollution Fee System

Since the pollution fees were introduced:

(a) There has been a strong incentive to monitor the development of both the system and emissions.

(b) The number of enterprises with agreed maximum permissible levels and temporary coordinated standards of emissions has risen sharply. The fee system stimulated enterprises increasingly to sign such agreements. The results of the two-rate system of pollution fees were significant since emissions of all enterprises who had not signed agreements, and had no fixed TCS, were regarded as being emissions above the standards and hence subject to payment of the highest fee level. Thus enterprises had a strong incentive to collect documents to determine their permitted levels of emissions.

(c) The increased accumulation of environmental funds promoted the development of markets for ecological services, although it was hard to spend money because of the narrow market for abatement equipment; procedures for decision making on resource allocations were not sufficiently advanced; and conflicts arose between executive and legislative authorities as to the directions of resource allocations.

Our analysis has shown that the majority of environmental expenditures have been made in the regions of the experiment participants (three-quarters of them spent no less than 80% of their own resources in 1991). This in itself is testimony to the high capacity of the environmental authorities to learn and to adapt the environmental management system very quickly to the new conditions of the forming market.

To make some quantitative estimates, let us consider the dynamics of air pollution from stationary sources in Russia (*Figure 8.1*). One can see that there was a constant decrease in harmful emissions until 1989. In 1990 there was some increase in emissions, probably due to improvements in monitoring systems and to more accurate reporting as a result of the experiment with the introduction of pollution fees in some regions. The necessity to calculate fees stimulated more detailed emissions accounting.

From 1991 to 1993 there was a sharp decline in emissions, at the highest rates for the last ten years. This could be evidence for the efficiency of the pollution fee system, but we must also take into account the overall economic situation in Russia in this period. There were sharp reductions in all production indices. *Figure 8.2* presents the dynamics of emissions per unit of GNP in Russia. Although there has been a reduction in total emissions, the amount per unit of GNP has actually increased. Thus production processes appear to have become more environmentally intensive, and the pollution fee system is now less effective than the CAC policy.

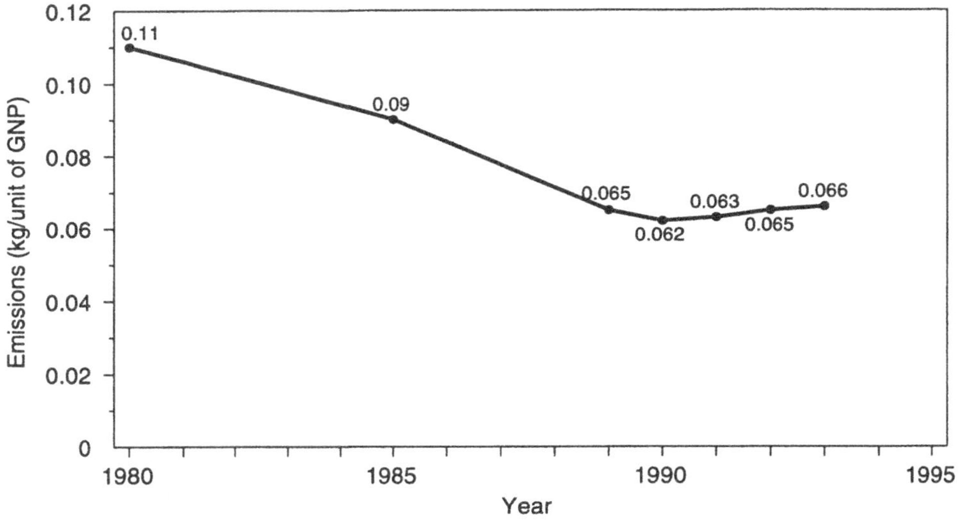

Figure 8.2. Dynamics of air pollution from stationary sources per unit of GNP. Source: Calculated according to USSR Statistical Committee, 1985, 1988, 1989, 1990.

8.4.1 Development of pollution investments

It is very useful to analyze the dynamics of investments in pollution abatement technologies. We consider the share of abatement investments in GNP, since the high rates of inflation make it impossible to analyze absolute values (*Figure 8.3*). The share of abatement investments in GNP has fallen sharply since 1980 and they now constitute just 0.12% of GNP, compared with 0.35% in 1989. This is new evidence of the rather low effectiveness of the pollution fee system in maintaining the necessary quality of the environment. We also show in *Figure 8.3* the share of abatement investments in GNP needed to reach the environmental standards of the State Long-Term Environment Program by the year 2005.

A number of factors can explain the poor performance of the new environmental management system:

- During the economic crisis the allocation of financial resources from the state budget for environmental protection almost stopped.
- Environmental funds and the enterprises' own resources (considered as part of pollution fees) became the only sources of finance for investments in pollution abatement.

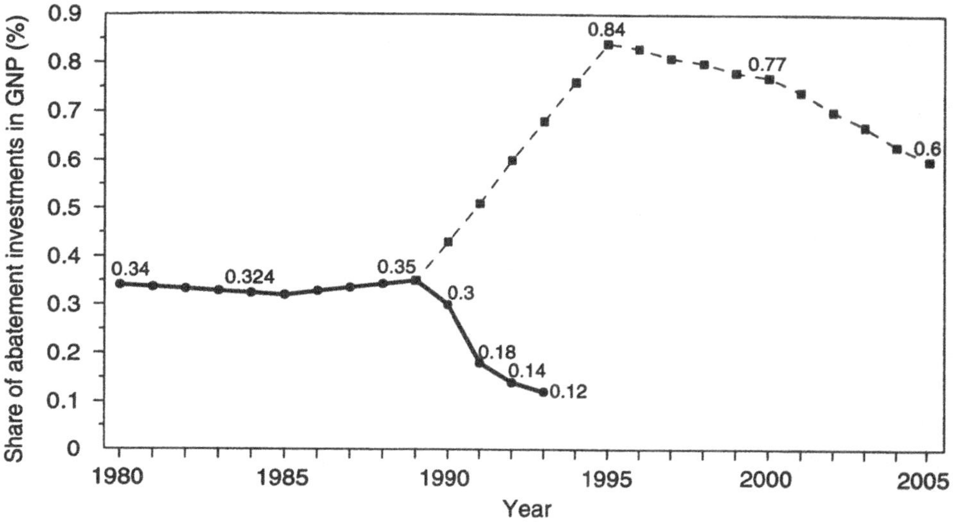

Figure 8.3. Dynamics of the share of abatement expenditures in GNP. Full curve, actual dynamics, 1980–1993; dotted curve, share of abatement investments in GNP required to meet environmental goals by the year 2005 according to the LTEP (USSR State Committee on Scientific and Technological Development, 1980). Source: Calculated according to USSR Statistical Committee, 1985, 1988, 1989, 1990.

- The only opportunity to increase the share of abatement expenditure in GNP to 0.35% or to the more desirable level of 0.8% is to increase the rates of payments by 3 or 7 times, respectively. Then appropriate resources would accumulate in environmental funds, to be spent for abatement purposes.

Environmental funds are now the main source of abatement investments, but because of the low fee rates these funds lack sufficient resources. Moreover, the funds are unable to collect all the necessary payments. Today, another reason is dominant: whereas in 1991 the funds lacked the legal authority to collect payments from enterprises, the main reason now is that most enterprises are either insolvent or on the verge of bankruptcy. In the first quarter of 1993 only 30% of the necessary fees were paid. Enterprises managed to cancel their debts only after obtaining large-scale credit from the Russian Central Bank. In 1994 the situation is continuing as high rates of inflation devalue the accumulated resources even further.

We have not mentioned the abatement investments made by enterprises from their own resources. If the nominal interest rate is more than 200% (the real interest rate is negative), only very profitable investments are justified. Enterprises would use their resources for abatement investments if it would be more profitable than to pay pollution fees. On the basis of case studies and analyses of branches of industry, we conclude that in the current situation fees are not providing sufficient motivation for enterprises to use their own resources for environmental protection.

The same results were obtained from a survey of 65 large enterprises in 24 regions (Dumnov, 1992). Dumnov calculated that the average enterprise spent 0.2% of total annual receipts on pollution fees. In general, pollution payments had little influence, but 80% of the managers of enterprises surveyed reported that they had increased the prices of their products to compensate for the fees. If all the levied pollution fees had actually been collected, the amount received would have been 70% higher than that actually realized. No one spoke about the necessity for new environmental investments. If in 1991 only 8% of local enterprises were stimulated by the pollution fees, now it is less than 1%. Pollution fees will create more incentives for abatement investments when the Russian economy perceives market signals.

To observe one more reason for environmental degradation, we compare the dynamics of abatement investments with all investments in Russia. The former have been decreasing at a higher rate than the latter. Of course, the economic crisis has influenced the pattern of expenditures, but even in comparison with the dynamics of all investments (which fell by a factor of 2–2.5; Shatalin, 1994), environmental protection investments fell even further (by a factor of 4–5). The indexation of pollution fees at the beginning of 1993 improved the situation, but only temporarily.

8.5 Conclusion

The main conclusion from this chapter is that the environmental sector in Russia has been more sensitive to the general deterioration of the economic situation than other sectors. The creation in recent years of the pollution fee system was unable to prevent a decrease in environmental protection activity that has been sharper than in other sectors.

The transition from the command-and-control system to the pollution fee system seems to be a natural process for Russia as the country moves toward a more market-oriented economy. Our results, however, suggest that the introduction of pollution fees in a period of transition has to be accompanied by appropriate measures that mitigate the negative consequences of the current economic crisis on the development of the environmental sector. The introduction of the system of pollution fees did create the institutional foundations for a new environmental management system, and we expect this system to become more efficient as the market economy in Russia develops.

Note

[1] For example, the MPC for dust emissions was exceeded more than 3 times in 16 towns (by factors of 6 in Donetsk and Osh; 5.3 in Frunze and Rustavi; 4 in Fergana; 3.4 in Dnepropetrovsk, Odessa, and Makeevka). For SO_2, emissions standards were exceeded by factors of more than 3 in 14 cities (Astrakhan 6.2; Kirovakan 4.8; Krivoy Rog 4.4; Norilsk and Novotroitsk 4; Grozny 3.8; Cheljabinsk, Saratov, Jaroslavl, and Donetsk 3).

References

Dumnov, A., 1992, Economic Methods of Environmental Protection and Arrangement of Statistical Surveys, *Vestnik Statistiky* 5:55–57.

Gofman, K., 1977, *Estimates of the Economic Value of Natural Resources*, Nauka, Moscow, Russia.

Gofman, K.G., and Gusev, A.A., 1977, *Environmental Protection: Models of Environmental Quality Control*, Economica, Moscow, Russia.

Gofman, K.G., Gusev, A.A., and Balatsky, O.F., 1986, *Temporary Rules for Evaluating the Effectiveness of Environmental Measures and Estimating Economic Damage*, Economica, Moscow, Russia.

Russian Council of Ministers, 1991, *The Polluting Emissions Fee System and Its Implementation*, (13), Russian Council of Ministers, Moscow, Russia.

Russian Council of Ministers, 1992, *The Adoption of Rules for Calculating Pollution Charges and Their Marginal Level*, (632), Russian Council of Ministers, Moscow, Russia.

Russian Supreme Soviet, 1991, *The Basic Taxation System Act of the Russian Federation*, Politizdat, Moscow, Russia.

Russian Supreme Soviet, 1992, *Environmental Protection Act*, Supreme Soviet, Moscow, Russia.

Shatalin, S., 1994, Market Needs Management, *Economics and Life* 5:1.

USSR Council of Ministers, 1978, *Additional Measures for Environmental Protection and the Rational Utilization of Natural Resources*, Decisions of the USSR Council of Ministers, (96), Politizdat, Moscow, Russia.

USSR Council of Ministers, 1980, *The MPLs of Pollutants into the Atmosphere and Their Harmful Physical Influences*, Decisions of the USSR Council of Ministers (4), Article 18, Politizdat, Moscow, Russia.

USSR State Committee on the Scientific and Technological Development, 1980, *The Long-Term Program for Environmental Protection and the Rational Utilization of Natural Resources up to 2005*, GKNT, Moscow, Russia.
USSR Statistical Committee, 1985, *National Economy of the USSR in 1985*, Finance and Statistics, Moscow, Russia.
USSR Statistical Committee, 1988, *National Economy of the USSR in 1988*, Finance and Statistics, Moscow, Russia.
USSR Statistical Committee, 1990, *National Economy of the USSR in 1990*, Finance and Statistics, Moscow, Russia.
USSR Statistical Committee, 1989, *Environmental Protection and Rational Utilization of Natural Resources*, Finance and Statistics, Moscow, Russia.
USSR Statistical Committee, 1990, *Environmental Protection and Rational Utilization of Natural Resources*, Finance and Statistics, Moscow, Russia.
USSR Supreme Soviet, 1968, *The Basic Land Management Act in the USSR and Soviet Republics*, Politizdat, Moscow, Russia.
USSR Supreme Soviet, 1969, *The Basic Public Health Act of the USSR*, Politizdat, Moscow, Russia.
USSR Supreme Soviet, 1970, *The Basic Water Management Act in the USSR and Soviet Republics*, Politizdat, Moscow, Russia.
USSR Supreme Soviet, 1975, *The Basic Mining Management Act in the USSR and Soviet Republics*, Politizdat, Moscow, Russia.
USSR Supreme Soviet, 1977, *The Basic Forests Management Act in the USSR and Soviet Republics*, Politizdat, Moscow, Russia.
USSR Supreme Soviet, 1977, *The Constitution (Main Law) of the USSR*, Politizdat, Moscow, Russia.
USSR Supreme Soviet, 1980, *Atmospheric Protection Act*, Politizdat, Moscow, Russia.
USSR Supreme Soviet, 1980, *Wildlife Protection and Utilization Act*, Politizdat, Moscow, Russia.

Part III

International Aspects

Chapter 9

Making Carbon Emission Quota Agreements More Efficient: Joint Implementation versus Quota Tradability

Peter Bohm
Department of Economics
Stockholm University, Stockholm, Sweden

Abstract

Given that a set of countries agree to keep their carbon emissions within certain national quota limits, the outcome would be a treaty in the tradition of the Rio Convention, only one step more demanding. This chapter discusses two additional measures to improve the efficiency properties of such a nontradable quota treaty. Both measures would allow committed countries to buy emission reductions from other countries as a way of partially fulfilling their commitments. One is "joint implementation", which was supported by the Convention as an instrument capable of increasing the efficiency of commitments to limit carbon emissions. It is questioned here whether joint implementation between developed and developing countries would imply a considerable increase in efficiency. The other measure is to make national emission quotas tradable, which, in contrast, could significantly improve efficiency (to the extent that trade does), especially if developing countries with low abatement costs could be induced to participate. Here, we identify the circumstances under which quota tradability would allow developing countries to participate without subjecting them to risks of economic losses.

Key words: carbon abatement policy, carbon abatement costs, joint implementation; tradable permits.

9.1 Introduction

If a significant number of governments continue their efforts to reach an agreement on carbon emission reductions, their policy considerations for the short run are likely to differ from those for the long run. The short-run objectives will probably center on getting started on a noticeable scale with respect to the number of signatories, but with modestly ambitious reduction goals. The Framework Convention on Climate Change (FCCC) of 1992 may be interpreted as such an attempt. The long-run objective will probably be somewhat different and focus on attaining significant reductions in global emissions among those countries that agree to cooperate. This means that the long-run objectives would center on attaining an efficient international abatement policy, i.e., resulting in a maximum carbon emission reduction for a given set of national sacrifices. In this chapter, we assume that, in the near future, the aggregate international benefits of greenhouse gas or carbon emission reductions cannot be determined with any great degree of precision. As a result, an *optimal* level of emissions could not be estimated. Instead, it is postulated that the target for an agreement among a set of countries will be cast in terms of staying within a certain, more or less arbitrarily chosen, level of aggregate emissions. (*Efficiency* under these circumstances means *cost effectiveness*; these two terms are used interchangeably here.)

The FCCC does not formally require an individual signatory country to reduce its greenhouse gas emissions to any specific level. However, the Convention requires that Annex I signatories (essentially developed countries) "communicate ... detailed information on their policies and measures ... with the aim of returning individually or jointly to their 1990 levels", by the year 2000, of emissions of CO_2 and other GHGs (not controlled by the Montreal Protocol on Substances that Deplete the Ozone Layer). If, as a result of the FCCC, these countries do return individually to their 1990 emission levels by the year 2000, this would represent an internationally inefficient carbon abatement solution, similar to that of the Montreal Protocol, where the signatories' emission quotas were not set so as to minimize abatement costs for the set of signatory countries. [On the efficiency implications of the Montreal Protocol, in general, and its set of nontradable chlorofluorocarbons (CFC) emission quotas, in particular; see Bohm, 1990.] However, the FCCC allows parties to engage in joint implementation as a means of meeting the commitments of individual signatories by financing emission reductions in other countries. Joint implementation has been hailed as an important step toward efficiency; see, for example, Barrett (1993a, 1993b) and Jones (1993), and possibly as a first practical step toward an international tradable quota treaty, which under conditions of a perfect quota trade market would represent an efficient long-term carbon policy. A tradable quota treaty is one where signatories have fixed initial carbon emission quotas and can trade parts of these quotas. Joint implementation among countries with commitments to limit their carbon emissions could develop into a full-scale tradable quota system. In fact, this type of joint implementation could hardly be anything less than a version of such a system, since the countries involved would have to prove that they have reached a level of joint emissions not exceeding the sum of their emission commitments (see Bohm, 1994).

In this chapter, joint implementation (JI) refers only to cases where a country with a fixed emission quota, e.g., an industrial country that (eventually) aims to keep its emissions in the year 2000 below its 1990 level, is credited for financing emission reductions in a country without any emission reduction commitment, here taken to be a developing country. More specifically, we assume that there is a set of developed country FCCC signatories who (at some future date) have made commitments in terms of (binding) national emission quotas and who may buy emission reductions from poorer countries, (either nonsignatories or signatories who have not made any emission commitments) as a means of staying within their national quotas. Obviously, this case is where JI introduces something new. This is also the case that has been in focus in the rapidly growing (but so far mainly unpublished) literature on JI and that will be dealt with here.

Two issues are discussed in this chapter. First, it is questioned whether JI with developing countries in practice can significantly improve efficiency in a nontradable quota treaty among industrial countries, and whether JI would help materialize a future tradable quota (TQ) treaty in which developing countries would also participate. Second, we investigate the distributional implications of making national emission quotas tradable, particularly the effects on developing countries whose participation is so vital for making a carbon treaty efficient. This shift from inefficiency to efficiency is essentially a consequence of significant differences in marginal abatement costs across countries, which a TQ treaty would tend to equalize.

Of course, efficiency could also be accomplished through a "global" carbon tax or a system of internationally harmonized domestic carbon taxes. Before proceeding to a discussion of these issues, it is explained (in Section 9.2) why neither an international tax, i.e., where tax revenues are collected by an international organization, nor a system of harmonized domestic taxes is likely to be a successful candidate for an optimal long-term policy.[1] In Section 9.3, we briefly discuss the extent to which a system of bilateral JI between industrial countries and developing countries is capable of significantly improving efficiency and of developing into a tradable quota system including also developing countries. In Section 9.4, the main part of the chapter, we investigate the effects of making national quotas tradable on different types of countries, assuming that developing countries would also participate. We assume that, domestically, all countries use some efficient policy instrument to stay within their quotas, be it a domestic tradable permit system, a domestic carbon tax, or some other approach. These effects essentially turn on two factors: initial quota allocations to countries, and the levels of their abatement costs. In both these respects, developed and developing country parties to an international carbon reduction treaty are likely to differ in a systematic fashion.

9.2 Instruments for an International Carbon Emission Reduction Policy

A major reason why an *international carbon tax* can hardly be regarded as a feasible approach to a stringent carbon emission reductions policy is that many countries – especially those that use relatively large quantities of fossil fuel and would hence pay large amounts of carbon taxes – are unlikely to accept that "others" (the international organization) would be given full control over *their* tax revenues. In a more promising version of an international carbon tax, suggested by Hoel (1993), the percentages according to which the total tax revenues to the international organization would be shared among the signatories are taken to be determined at the outset. A problem with this version, however, is that even if the set of cosignatories and the tax level could be assumed to be given, the effect of the tax on the volume of total taxes from the cosignatories, and hence the effect on the amount of tax redistributions to the individual party, would not be known beforehand. This is likely to make many countries hesitate to join such a treaty.

A system of *harmonized carbon taxes* (and tax transfers required in order for poor countries to cooperate) is hardly likely to be sustainable, particularly if a number of smaller or poorer parties to such an agreement felt more or less forced to join the agreement. A reason for this feeling is the potential threat embodied in the arsenal of "carrots" and "sticks" (aid and trade policy instruments) that a strong subset of committed countries, such as the OECD, has at its disposal. If so, individual countries may (be expected to) do what they can to circumvent the effect of the carbon tax on their fossil fuel consumption. Obviously, a practicable agreement cannot be one that stops signatories from using all policy instruments with potential direct or indirect effects on fossil fuel consumption. Thus, a tax harmonization agreement may be impracticable either because it turns out to be impossible to control all instruments, other than the carbon tax, which influence fossil fuel consumption, or because attempts to implement such control would be so unpalatable to individual countries that they decide not to join or later defect from the agreement (see also Hoel, 1993).

If these contingencies mean that carbon tax harmonization runs the substantial risk of being ineffective or unstable in the long run, it becomes particularly important to analyze the properties of the alternative system of carbon emission quotas. Regardless of whether the quotas are tradable or nontradable, a reluctance to cooperate may threaten the effectiveness and stability of such a system as well, although perhaps not to the same extent. Here, the problem is not that it would be difficult to determine how much carbon the treaty actually allows the signatories to emit, which, in effect, is the problem of a tax agreement just hinted at. The problem would now be that signatories may try to conceal their actual levels of fossil fuel consumption. This monitoring problem may be severe for consumption of fossil fuels produced domestically or imported from nonsignatory countries, but hardly for fossil fuels traded between parties to the agreement where it is in the interest of the exporting party that the quantities *not* consumed in the country are registered.

If a quota system were found to be practicable from the point of view of monitoring, there are other issues, particularly those concerning the implications of incomplete quota agreements, that need to be analyzed. A nontradable quota (NTQ) system is the precedent of the Montreal Protocol (with respect to CFC consumption).[2] Such a system is inefficient and is likely to be very costly to the world economy when the issue is shifted from reducing the use of ozone-depleting substances such as CFCs and halons, to that of reducing the use of fossil fuel. Hence, a carbon emission quota system would probably have to be made tradable. Since governments may hesitate to accept a policy mechanism that has not yet been used internationally and may even be poorly understood, it seems imperative that more is made known about the characteristics of a TQ system and how individual parties would be affected by the choice of a TQ instead of an NTQ system. To the extent that JI represents the generic element of a TQ system or even the beginning of such a system, it is important to gain further insight into what JI could offer in terms of efficiency improvements on a NTQ treaty.

9.3 Efficiency Implications of Bilateral Joint Implementation

9.3.1 Introduction

The discussion here concerns JI between an industrialized and a developing country. In fact, it has been argued that JI may be limited to such pairs of countries (for an interpretation of JI in the FCCC, see Barrett (1993b). Since the exact definition and criteria for JI still remain to be seen, it is meaningful at this point to observe only certain general characteristics of JI. We briefly discuss the following three issues:

1. Can bilateral JI be efficient, e.g., in the sense that buyers pay only the net marginal costs of emission reductions in other countries?
2. Can contracted net emission reductions in another country (one without a commitment to limit carbon emissions) be reliably monitored?
3. Can a bilateral JI system, even if quite extensive, evolve into a TQ system?

Purchases of emission reductions in another country can be made by the government of the buyer country, either directly from the government of the country where the abatement activity is undertaken, or indirectly via an international organization such as the Global Environmental Facility, or directly by private firms, although monitored by governments. Norway and a few other countries have already made pioneering steps to undertake what may represent officially recognized JI projects in the future.

JI activities are typically considered to be undertaken on a fairly large-scale *project level*, e.g., buying carbon offsets by (co)financing (in the Norwegian case) high-efficiency lighting, coal-to-gas conversion, and energy-efficient housing projects (Barrett, 1993b). For reasons of imperfect monitoring or control similar to those mentioned in the discussion of tax harmonization above, JI does not seem to be suitable in the form of "purchasing" the effects of the introduction of a carbon tax in another country. Nor would JI, for reasons of transactions costs, involve (co)financing truly small projects. Therefore, JI is also interpreted here as concerning large specific investment projects in the seller country.

The principal advantages of JI, most often highlighted, are (a) promoting global cost efficiency by replacing high abatement cost projects in one country by low abatement cost projects in another; hence, (b) allowing national abatement goals to be more ambitious; and (c) improving the chances of achieving a more comprehensive international abatement agreement (see, in particular, Hanisch, 1991; Hanisch *et al.*, 1992; Barrett, 1993a, 1993b; Jones, 1993). Moreover, especially if JI activities were worldwide and quite extensive, JI could significantly reduce the carbon leakage problem which is a major complication for incomplete tax or quota treaties. Finally, given its project-by-project approach, JI would be relatively easy to start up; hence, it could at least be a short-term second-best solution or the only politically feasible solution in the short term.

However, as will be shown here, some aspects of this list may present too rosy a picture of the implications of allowing JI. In particular, it needs to be ascertained that JI does not lead up a blind alley, making it even more difficult to reach a significant international agreement to reduce greenhouse gas emissions. But, first, in what sense does JI actually promote efficiency and to what extent is it practicable? Before setting out, it should be noted that most of the quoted writers on JI acknowledge some, or perhaps most, of the problems to be mentioned here. A possible point of disagreement, however, concerns the severity of these problems.

9.3.2 Efficiency aspects

The *potential* efficiency implications are impressive, especially if JI can effectively survive the first small-scale phase. If only a few JI projects are attempted, it is widely recognized that the transactions costs per project are likely to be large. Furthermore, overvalued currencies in seller countries may constitute a threat to the efficiency of JI. More important, to the extent that a thin (initial) JI market means insufficient competition among sellers, there is a substantial risk that sellers will get significantly more than their net marginal costs covered. Another important efficiency aspect relates to the next problem: will there be a systematic tendency to register an overestimate of the effect on net carbon emissions in the seller country?

9.3.3 Monitoring aspects

A first problem here concerns the project's direct effects on carbon emissions. This problem may be illustrated by a case that has been mentioned as a potential officially recognizable JI project. Although this case does not represent a carbon emission reduction project that could easily be incorporated in a quota or a tax harmonization solution, it nevertheless reveals problems that could arise in other JI projects as well.

The case involves a private US firm, AES Corporation, which paid $2 million of a project cost of $14.5 million for planting more than 50 million trees in Guatemala over a ten-year period, to reforest an area of 385 square miles (see Barrett, 1993a). Assuming this JI project were to be credited to the USA, it would be in the interest of the firm that wants to achieve something by increasing sequestration through this project, to have more sequestration per dollar of investment officially registered rather than less. The US government, if party to an agreement to limit net carbon emissions, would prefer that one of its firms contributed to more sequestration rather than less; the more sequestration that could be registered, the less costly the buyer country's own domestic carbon reductions would be. Finally, the party – government and/or firm – implementing the tree-planting operations in the seller country would probably be able to get paid more for (appearing to be) doing more. This would be so even if the payment equaled "incremental costs" only, since in any cost concept relevant here some minimum profit would have to be included; otherwise neither the recipient firm nor its government would be interested in cooperating, and most likely the more profits that could be included, the bigger the project's registered sequestration effect would be.

Now, since all parties are interested in having a given project look as productive as possible, they may do whatever they can to provide documentation that supports their interests and to avoid providing information to the contrary. First, the tree-growing capacity of the area to be reforested could be exaggerated. Second, the quality of the seedlings and planting work could be exaggerated. These problems could be at least partly avoided by measuring, not "intended" reforestation, but actual reforestation at some (much) later stage. For this monitoring to be satisfactory, it would probably be quite expensive. Furthermore, knowing the actual effects only at some time in the future may not prove satisfactory.

The most difficult monitoring problem, of course, is to estimate the sequestration level in the case that no project is undertaken. First, would there be no naturally growing vegetation if the project were not carried out? Or is there any other use of the land, labor and possibly domestic capital tied up in the present project that would be worthwhile from a sequestration point of view in the absence of the project and, if so, how would that affect net carbon emissions? Moreover – moving from direct project effects to indirect effects – if the project had not been carried out, would, as a result, some similar reforestation project be set up elsewhere in the country? If so, how accurately could the net effects then be estimated?

Although it is an obvious problem that the incremental abatement of a project cannot be measured with accuracy, the problem is compounded by the fact highlighted above that all parties directly involved in the project have a common interest in making the estimates look "good".

If all these estimation problems would disqualify projects of the type now discussed, practicable JI schemes may be quite limited and concern only a part of all those projects that would be undertaken if a tradable quota agreement existed, which directly or indirectly introduced a permit price or (domestic) tax on carbon emissions. This would, of course, constitute another efficiency problem.

9.3.4 The long-term implications of joint implementation

It has been suggested that JI as between developed and developing countries is quite similar to a TQ approach. First, however, it should be noted that TQ systems could realistically also cover measures to increase carbon sequestration only if total national carbon sequestration levels could be monitored sufficiently accurately, most likely by satellite inspection (see Dudek and LeBlanc, 1992). To simplify a comparison of JI and TQs here, let us assume that such monitoring will in fact turn out to be possible. Then, the question can be raised as to whether a comprehensive carbon emission TQ system, including developing countries that have committed themselves to certain net emission limits, can be expected to grow out of a system of JI conducted on a project-by-project basis.

Although there are obvious similarities, JI and TQs differ in several important respects. As mentioned earlier, some abatement activities, "small projects" and (other) effects of levying a tax or (permit) price on fossil fuel in the countries involved cannot be included in a JI scheme, but could be, if a TQ system existed. In addition, although negotiations over initial quotas in a TQ treaty can certainly be expected to be complicated by conflicting estimates of future baseline emissions in the various countries, quota trade – *once such a treaty is attained* – would be relieved of any consideration of baseline scenarios, which was shown above to jeopardize the practicability of a JI system.

Given this fundamental difference between a JI system and a TQ system in operation (the former is based on the estimated emission reduction from an unobservable baseline scenario for each project, and the latter is based on reductions of a national emission quota in one country being purchased by another country), there does not seem to be any natural path along that a JI system could develop into a TQ system. In fact, it seems easier to find circumstances that would present obstacles for such a development. Specifically, if a developing country already benefits from a number of JI activities, it may not find the remaining *additional* benefits from joining a TQ treaty convincing enough, whereas without a past history of benefiting from JI operations, it would have found the benefits large enough to have joined the treaty (see Bohm, 1994).

To sum up, it is obviously an easy task to find faults in a JI system, but there remains one advantage that the more ambitious alternatives, at the present time, do not seem capable of matching: the principle of JI has been shown to be politically acceptable. On the other hand, there is still the question of whether these alternatives and their implications are correctly understood. In the next section, we address a crucial distributional aspect of changing a nontradable quota system into a tradable one.

9.4 Distributional Effects of Quota Tradability

9.4.1 Introduction

Just as JI is interesting as a means of increasing efficiency by engaging low-cost developing countries (DCs) in global emission abatement, it is interesting to investigate whether in general the introduction of quota tradability would attract DCs to cooperate with those richer countries which have committed themselves to reducing their emissions below baseline. In this section, we analyze in particular the effects of quota tradability on the benefits to DCs under various international differences in abatement costs and initial quota allocations. The following specific issues are discussed:

- What are the implications of abatement cost differences for DC tradability gains?
- Would industrialized countries (ICs) take more advantage of DCs in any relevant sense, the relatively higher the abatement costs are in the ICs? (This question relates in particular to a point often made by politicians that TQ systems would be unfair to DCs.)
- If a DC turns out to value its fossil fuel use more than was initially estimated to be the case, will it lose from having signed a TQ treaty, whereby it will be an exporter of carbon abatement?

In Section 9.4.2, an attempt is made to identify a shortlist of quota allocation principles that are likely to be relevant for ICs versus DCs in an international TQ treaty. These allocation principles, which amount to assigning all costs to ICs while allowing DC emissions or DC GDP to be fully protected, are used in Section 9.4.3 to analyze how the tradability gains for ICs and, in particular, DCs are affected by systematic differences in abatement cost functions and by changes in these functions.

9.4.2 Basic assumptions concerning cost functions and initial quota allocations

There has been a long discussion in the literature about potentially fair principles for initial quota allocations (see Grubb, 1989; Barrett, 1992). The discussion has suggested principles such as (current or past) population size and GDP, with or without special considerations of the individual country's share of fossil fuel exports in GDP, basic heating/cooling "needs", or the energy demand structure of existing industry, etc. Although it has been difficult to identify criteria that are likely to be widely accepted, it seems possible to suggest a shortlist of more generic allocation principles that will allow a theoretical analysis of the issues spelled out above.

To begin with, there seem to be two types of potential carbon emission treaty signatories: (a) those who require (at least) full compensation for all foreseeable implications of the treaty they are asked to join, and (b) those who accept to make sacrifices. The latter group can be divided into two subgroups: (b.1) those countries that agree to make sacrifices, while allowing certain other countries to be fully compensated, and (b.2) those that require all parties to make (some) sacrifices. This provides a set of two versions of an initial incomplete treaty, (I) to include only countries of type b.2 (which all agree to make sacrifices), and (II) to include countries of type a and b.1. These two versions can be interpreted as (I) a treaty among some or all ICs only, and (II) a treaty among some DCs and some, or most, of the ICs. Here, we focus on allocation principles for the perhaps more interesting case of a type II treaty, since it involves a broader spectrum of countries, including also DCs.

Then, a treaty of type II would be one in which DCs are (at least) fully compensated (as in the initial phase of the Montreal Protocol). In Bohm and Larsen (1993), two versions of this principle are used and will be used here as well: (1) where DCs are allowed to remain on their estimated baseline (or "business-as-usual") emissions path, which would benefit them if a quota trade were to open up, and (2) where the allocations are estimated to keep DCs fully compensated (plus some minimal incentive to make them prefer to join the treaty). Obviously, for reasons mentioned in connection with our discussion of JI, these estimates will not be easily agreed upon. Still, it is assumed here that, given negotiations setting out from any of these two principles, some decision on emission levels for a set of DCs would eventually

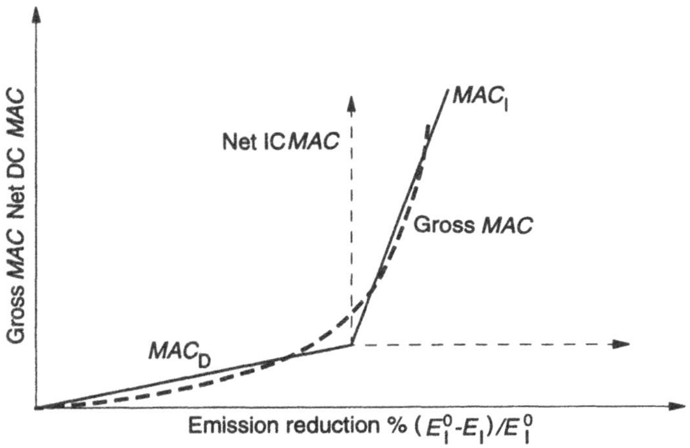

Figure 9.1. Schematic marginal abatement costs for an industrial country and a developing country.

be arrived at, and henceforth their emissions would have to stay within the (post-quota trade) levels.

As already mentioned, all costs associated with treaties to be discussed here will be borne by the ICs. Although our subsequent analysis does not require any specific burden-sharing principle among the ICs, we may assume, to fix ideas, that a fair allocation among the countries which share the costs is one where costs amount to the same percentage of GDP (Bohm and Larsen, 1994).

The incomplete treaties to be discussed are all such that carbon tariffs or similar actions are taken by the signatories to prevent any significant carbon leakage through trade with nonsignatories.

Gross abatement cost functions may generally be seen as setting out from zero and having marginal abatement costs (*MAC*) grow at an increasing rate (see Nordhaus, 1991; Bohm and Larsen, 1994). This means that we disregard the existence of significant "no-regret" (i.e., costless) options (see the "gross *MAC*" curve in *Figure 9.1*).

We assume that ICs have moved farther up along this *gross MAC* curve than DCs, typically because fossil fuel use has been reduced more in ICs, mainly due to their more stringent actions to reduce domestic pollution from SO_x and NO_x emissions from fossil fuel combustion. Then, the *net MAC* curve (deducting the national valuations of domestic pollution) will be steeper for ICs than for DCs. Since the main cost differences between

DCs and ICs can be found in the relative slopes of the *MAC* functions, we use linear (net) *MAC* functions with the slope being steeper for ICs than for DCs; see the net *MAC* curve for a typical DC (MAC_D) and the net *MAC* curve for a typical IC (MAC_I) in *Figure 9.1*.

9.4.3 Implications of differences in MACs and initial quota allocations

We set out from a nontradable quota treaty where each IC is constrained by an emission quota (which, for simplicity, is assumed to be constant over time), while DCs, being nonsignatories, can emit carbon as they please. We compare this treaty with two extreme cases of a tradable quota treaty in which also DCs participate and where, in case 1, their initial quota allocations equal what they are expected to emit as a maximum under "business-as-usual" and, in case 2, their initial allocations are such that they would be fully compensated for their emission commitments.

Without any loss of generality, we limit the discussion to a treaty between two signatory countries, an IC (indexed I) and a DC (indexed D). They are assumed to behave as if there were many signatories, where no country has any significant effect on quota or permit prices p. Country I has agreed to stay within its quota allocation $E_{\mathrm{I}}^{\mathrm{A}} < E_{\mathrm{I}}^{0}$, its (estimated) baseline emissions.

Case 1

Country D emits carbon equal to its projected maximum baseline emissions; thus, as an allocation in a TQ treaty, we would have $E_{\mathrm{D}}^{\mathrm{A}} = E_{\mathrm{D}}^{0}$. Then, total emissions, E^{T}, from the two countries are here taken to be:

$$E^{\mathrm{T}} = E_{\mathrm{I}}^{\mathrm{A}} + E_{\mathrm{D}}^{\mathrm{A}} < E_{\mathrm{I}}^{0} + E_{\mathrm{D}}^{0}.$$

To simplify the arithmetic and diagrammatic presentation, we assume that $E_{\mathrm{I}}^{0} = E_{\mathrm{D}}^{0}$. Then, the net abatement cost function and the net *MAC* function for country i can be written, in absolute terms, as

$$C_i = 1/2\ c_i(E_i^0 - E_i)^2; \quad MAC_i = c_i(E_i^0 - E_i),$$

where E_i denotes actual emissions (see *Figure 9.1*). The total costs of a nontradable quota (NTQ) and a tradable quota (TQ) treaty are

$$C^{\mathrm{NTQ}} = 1/2\ c_I(E_{\mathrm{I}}^{0} - E_{\mathrm{I}}^{\mathrm{A}})^2,$$

$$C^{\mathrm{TQ}} = 1/2\ c_I(E_{\mathrm{I}}^{0} - E_{\mathrm{I}}^{*})^2 + 1/2\ c_{\mathrm{D}}(E_{\mathrm{D}}^{0} - E_{\mathrm{D}}^{*})^2,$$

where E_i^* denotes equilibrium emissions after quota-unit trade. Thus, E_i^* and quota-unit price p are determined by the following set of equations, where equation (9.1) shows the equality in equilibrium between the MAC_i's and p, using the fact that quota exports $(E_D^0 - E_D^*) = (E_I^* - E_I^A)$, quota imports.

$$c_I(E_I^0 - E_I^*) = c_D(E_D^0 - E_D^*) = c_D(E_I^* - E_I^A) = p, \tag{9.1}$$

$$E_I^* + E_D^* = E^T.$$

This yields

$$E_I^* = \frac{c_I E_I^0 + c_D E_I^A}{c_I + c_D}. \tag{9.2}$$

Aggregate tradability gain G is defined as

$$G = C^{NTQ} - C^{TQ},$$

which can be written

$$G = \frac{1}{2}\left[c_I(E_I^0 - E_I^A)^2 - c_D(E_D^0 - E_D^*)^2 - c_I(E_I^0 - E_I^*)^2\right]. \tag{9.3}$$

Replacing $c_D(E_D^0 - E_D^*)$ by $c_I(E_I^0 - E_I^*)$, (see equation (9.1), and $(E_D^0 - E_D^*)$ by $(E_I^* - E_I^A)$, the second term becomes $-c_I(E_I^0 - E_I^*)(E_I^* - E_I^A)$. Hence, we obtain

$$G = \frac{1}{2}\, c_I(E_I^0 - E_I^A)(E_I^* - E_I^A) > 0, \tag{9.4}$$

since, by definition,

$$E_I^0 > E_I^* > E_I^A$$

for an IC signatory. That $G > 0$ is, of course, the obvious effect of opening up a new kind of trade. The gain is shown by the shaded triangle in the box diagram of *Figure 9.2*, where emissions from country I are read from left to right and emissions from country D are read in the opposite direction; the reduction in total emissions is shown by shrinking the box by $E_I^0 - E_I^A$, making the MAC_I and MAC_D curves between E_I^0 and E_I^A reflect quota demand and supply, respectively, on the international quota trade market.

The effect on G of increased marginal abatement costs for country I, given by an increase in c_I, equals

$$\frac{\partial G}{\partial c_I} = \frac{1}{2}(E_I^0 - E_I^A)\left[(E_I^* - E_I^A) + c_I\frac{\partial E_I^*}{\partial c_I}\right] > 0, \tag{9.5}$$

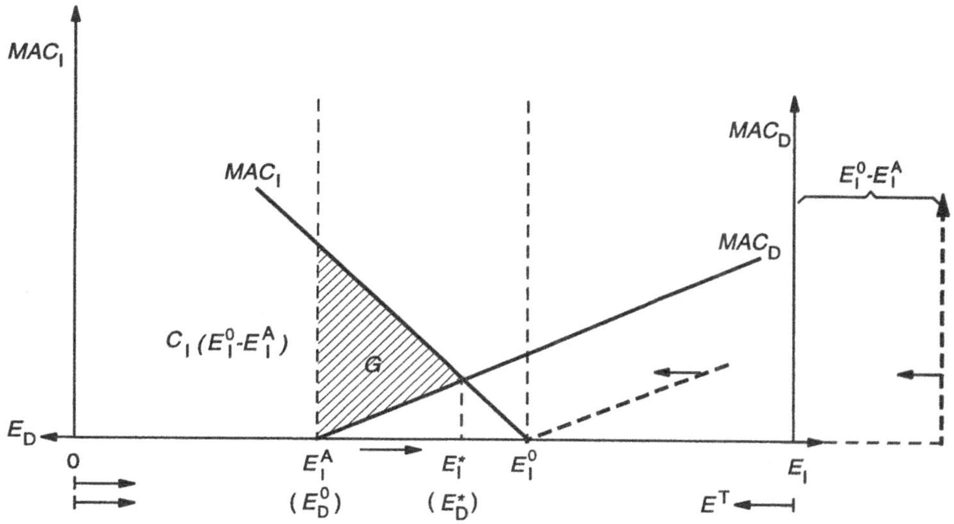

Figure 9.2. Tradable quotas for an industrial country and a developing country.

since

$$\frac{\partial E_\mathrm{I}^*}{\partial c_\mathrm{I}} = \frac{c_\mathrm{D}(E_\mathrm{I}^0 - E_\mathrm{I}^\mathrm{A})}{(c_\mathrm{I} + c_\mathrm{D})^2} > 0.$$

This means that increased *MACs* for an IC must increase the gains from making quotas tradable. Diagrammatically, this can be seen from *Figure 9.3a*, where a higher c_I means that both the base and the height of the G triangle increase when c_I increases. Setting out from a case where $c_\mathrm{I} = c_\mathrm{D}$, equation (9.5) also shows (a) the effect of making c_I larger than c_D and (b) the effect of further increases in c_I (e.g., when country I takes actions for domestic reasons that indirectly reduce carbon emissions).

Obviously, both I and D gain from opening up the quota trade, i.e., $G_i > 0 (i = \mathrm{I}, \mathrm{D})$. Net gains for D equals D's quota sales $p(E_\mathrm{I}^* - E_\mathrm{I}^\mathrm{A})$ minus abatement costs; i.e., using (9.1), we have

$$\begin{aligned} G_\mathrm{D} &= \underbrace{(E_\mathrm{I}^* - E_\mathrm{I}^\mathrm{A})}_{\text{quota sales}}\underbrace{c_\mathrm{D}(E_\mathrm{I}^* - E_\mathrm{I}^\mathrm{A})}_{\text{quota price}} - \underbrace{\frac{1}{2}c_\mathrm{D}(E_\mathrm{I}^* - E_\mathrm{I}^\mathrm{A})^2}_{\text{abatement costs}} \\ &= \frac{1}{2}\, c_\mathrm{D}(E_\mathrm{I}^* - E_\mathrm{I}^\mathrm{A})^2 > 0. \end{aligned} \qquad (9.6)$$

Using equation (9.1), we may rewrite equation (9.6) as

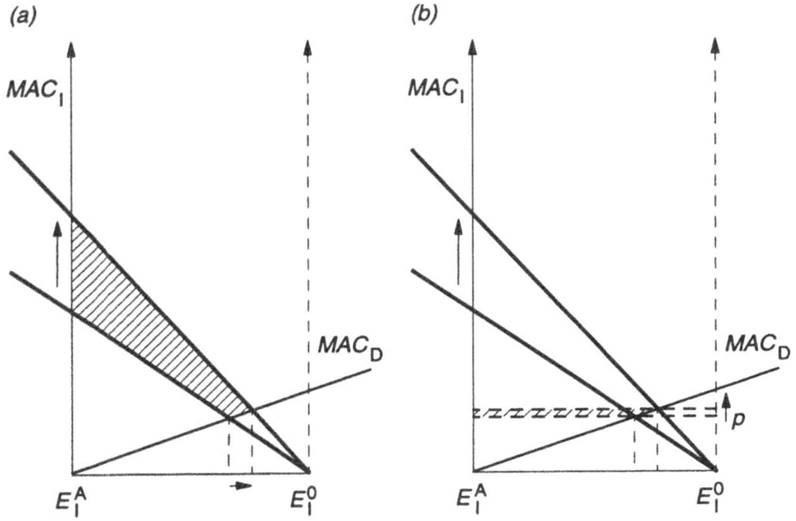

Figure 9.3. Effects of increased marginal abatement costs on tradability gains.

$$G_D = c_I(E_I^* - E_I^A)(E_I^0 - E_I^*) > 0. \tag{9.6'}$$

Similarly, using equations (9.4) and (9.6′), we have country I's gain as

$$\begin{aligned} G_I &= G - G_D \\ &= \frac{1}{2}\, c_I(E_I^0 - E_I^A)(E_I^* - E_I^A) - \frac{1}{2}\, c_I(E_I^* - E_I^A)(E_I^0 - E_I^*) \\ &= \frac{1}{2}\, c_I(E_I^* - E_I^A)^2 > 0. \end{aligned} \tag{9.7}$$

Thus, $G_I \geq G_D$ for $c_I \geq c_D$. Since

$$\frac{\partial E_I^*}{\partial c_I} = \frac{c_D(E_I^0 - E_I^A)}{(c_I + c_D)^2} > 0,$$

we also have that G_D grows with c_I, as illustrated in *Figure 9.3b*.

The effects of increases in c_D are less obvious for the case where c_D remains lower than c_I. Deriving equation (9.2) with respect to c_D, we get

$$\frac{\partial E_I^*}{\partial c_D} = \frac{c_I(E_I^A - E_I^0)}{(c_I + c_D)^2} < 0. \tag{9.8}$$

Thus, it follows from (9.4) and (9.7) that both

$$\frac{\partial G}{\partial c_D} < 0 \quad \text{and} \quad \frac{\partial G_I}{\partial c_D} < 0.$$

Table 9.1. Effects of increases in *MACs* where $E_{\rm I}^{\rm A} = E^{\rm T} - E_{\rm D}^{\rm A}$ and $c_{\rm I} > c_{\rm D}$. Case 1: $E_{\rm D}^{\rm A} = E_{\rm D}^{0}$ ($E^{\rm T}$ constant).

Effect of increase in cost ...	... on tradability gains G	=	$G_{\rm I}$	+	$G_{\rm D}$
$c_{\rm I}$	> 0		> 0		> 0
$c_{\rm D}$	< 0		< 0		> 0[a]

[a]Since $c_{\rm I} > c_{\rm D}$.

Given equations (9.6′) and (9.8), it looks as if $G_{\rm D}$ could either grow or fall with $c_{\rm D}$. To follow what happens in terms of *Figures 9.2* or *9.3*, we can note that $G_{\rm D}$ would obviously be at a maximum (for a given $c_{\rm I}$) if

$$E_{\rm I}^{*} - E_{\rm I}^{\rm A} = E_{\rm I}^{0} - E_{\rm I}^{*}.$$

This equality holds, as can be seen from equation (9.2), rewritten as $c_{\rm D}(E_{\rm I}^{*} - E_{\rm I}^{0}) = c_{\rm I}(E_{\rm I}^{0} - E_{\rm I}^{*})$, only if $c_{\rm D} = c_{\rm I}$. However, since we now have that $c_{\rm D} < c_{\rm I}$, equation (9.2) implies that

$$E_{\rm I}^{*} - E_{\rm I}^{\rm A} > E_{\rm I}^{0} - E_{\rm I}^{*}.$$

Hence, given equation (9.8), the effect of $c_{\rm D}$ on $G_{\rm D}$ has a unique and *positive* sign (which, of course, is what we would find from the first derivative of equation (9.6′) with respect to $c_{\rm D}$, although this is less transparent in terms of *Figures 9.2* and *9.3*). Thus,

$$\frac{\partial G_{\rm D}}{\partial c_{\rm D}} > 0.$$

To sum up, we have for case 1 that, given $c_{\rm D}$, the higher the costs in the ICs, the larger will be the gain for DCs (and ICs) from making quotas tradable; thus, DCs gain from the fact that abatement costs are higher in ICs. In addition, an increased value of fossil fuel use in a DC and, hence, an increase in *MACs* in such a country, given its estimated maximum baseline emissions, will further increase the DC's tradability gain, given that *MACs* are higher in ICs (see *Table 9.1*).

Case 2

In this case, we have a different allocation principle for a TQ treaty from that in case 1. Here, the initial allocation to country D is smaller than its estimated maximum baseline emissions, $E_{\rm D}^{0}$; it is small enough to be (approximately) equal to

$$E_{\rm D}^{\rm A} = \frac{1}{2}\,(E_{\rm D}^{0} - E_{\rm D}^{*}).$$

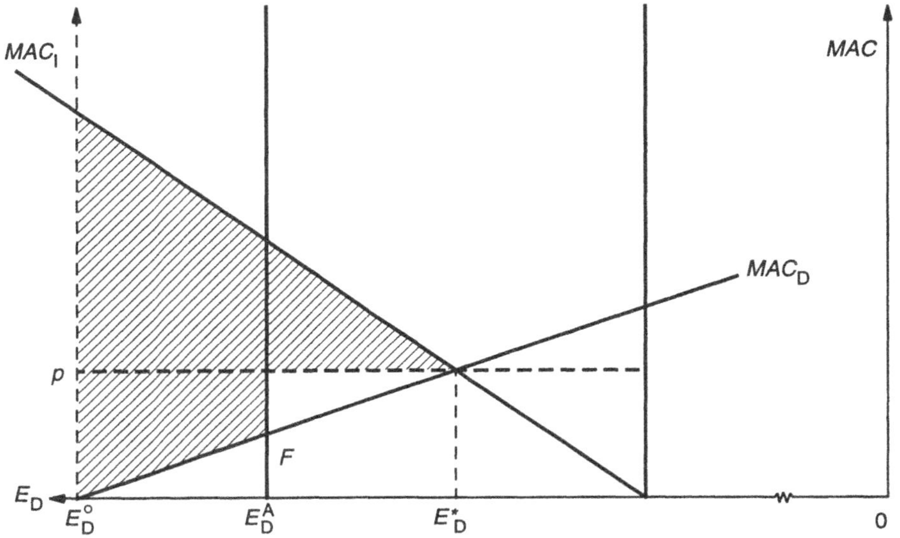

Figure 9.4. Tradability gain for an industrial country in case 2.

This means that, given (approximate) estimates of the equilibrium quota unit price and MAC_{D}, an initial allocation is sought where country D would be kept fully compensated, i.e., remaining on the same GDP level as in the case of a NTQ treaty (where $E_{\mathrm{D}} = E_{\mathrm{D}}^{0}$). (This can be seen in *Figure 9.4*, where the initial allocation $E_{\mathrm{D}}^{\mathrm{A}}$ implies that country D (a) has costs equal to area $E_{\mathrm{D}}^{0}E_{\mathrm{D}}^{\mathrm{A}}F$ for reducing its emissions to its initial allocation, and (b) receives compensating profits by selling emission reductions down to E_{D}^{*}.) Assuming that country D can be persuaded to join a treaty only if kept at least fully compensated, this case represents the opposite extreme to case 1. Here, all tradability gains would go to country I (which is the only party making a net payment), except perhaps for some (constant) minimum incentive premium to country D to make this county prefer to participate, as was suggested above.

C^{NTQ} is the same as before, since $E_{\mathrm{I}}^{\mathrm{A}}$ remains as in the previous case, and country D, by assumption, makes no emission reduction in a NTQ treaty. C^{TQ} and the size of the tradability gain going to country I depend on how the quota allocation to I is determined. Here, total emissions (E^{T}) remain the same as in case 1 and hence, $E_{\mathrm{I}}^{\mathrm{A}} = E^{\mathrm{T}} - E_{\mathrm{D}}^{\mathrm{A}}$. If so, country I would benefit from both an increased initial allocation and the tradability gain under the TQ system; see the shaded area in *Figure 9.4*. Since E^{T} is unchanged, the only difference from case 1 is that initial allocations between

I and D are shifted from E_{D}^{0} to $E_{\mathrm{D}}^{\mathrm{A}}$ in *Figure 9.4*. Obviously, the higher c_{I} is, the larger this gain will be. Hence,

$$\frac{\partial G}{\partial c_{\mathrm{I}}} = \frac{\partial G_{\mathrm{I}}}{\partial c_{\mathrm{I}}} > 0.$$

Similarly, it is obvious that

$$\frac{\partial G}{\partial c_{\mathrm{D}}} = \frac{\partial G_{\mathrm{I}}}{\partial c_{\mathrm{D}}} < 0.$$

In contrast with case 1, a distinction must now be made between *ex ante* and *ex post* changes in c_{I} and c_{D} for the case of country D. Obviously, by the construction of the allocation principle for this country, it will be kept fully compensated for *ex ante* changes in the *MACs*. For increases in *MAC ex post*, it can be seen from *Figure 9.4* that country D would gain from a larger c_{I} but lose from a larger c_{D}. This means that, for country D to be certain of being fully compensated, the allocation to D must be based on an upper bound to the estimated c_{D}.

More relevant, perhaps, is the case where country I's NTQ emission target, $E_{\mathrm{I}}^{\mathrm{A}}$, remains intact (case 2′). Since the effects on G_{I} can now be shown to be ambiguous (although always positive), let us concentrate on the effect on G_{D}, which is more important here. Since $E_{\mathrm{I}}^{\mathrm{A}}$ is kept intact, the MAC_{D} curve in *Figure 9.4* must be shifted to the left by an amount equal to $E_{\mathrm{D}}^{0} - E_{\mathrm{D}}^{\mathrm{A}}$ in equilibrium. For *ex ante* changes in c_{I} and c_{D}, country D will, by assumption, still be fully compensated. However, for *ex post* increases in *MACs*, the effects on G_{D} will be the same as in case 2; again, the initial allocation to country D is given and an increase in c_{I} (c_{D}) means an increase in quota demand (decrease in quota supply). Hence, as before, since an *ex post* increase in c_{D} would imply a loss for country D, the initial allocation to this country would have to be based on the upper bound to the estimates of c_{D}. Obviously, for similar reasons, initial allocations to country D – in case 2 as well as case 2′ – cannot be based on what might turn out to be an overestimate of c_{I}.

The conclusions for cases 2 and 2′ to be highlighted here are that, by construction of the allocation principles, DCs are kept fully compensated and, hence, their GDPs are not affected by a TQ treaty regardless of the *ex ante* estimates of *MACs*. In addition, we have seen that DCs would gain from *ex post* increases in c_{I}. However, since *ex post* increases in c_{D} would imply a loss for DCs, their initial allocations must be based on an upper bound to their estimated *MACs*. For similar reasons, the DC allocations must be based on a lower bound to the estimated IC *MACs*. It should be added that, in case 2′, total emissions will decrease and thus provide a net benefit to DCs to the extent that they value a decrease in total carbon emissions.

As a final remark, it should be noted that if, from the outset and before negotiations reached the point where initial allocations had been decided upon, an NTQ treaty was considered along with a TQ alternative, the latter would be to the advantage of DCs in still another respect. As shown by Bohm (1992), the fact that tradability increases the value of quotas for those signatories who would end up being exporters of carbon abatement, typically DCs, and reduces the costs of commitments for those who would make significant commitments, typically ICs, the former would now want larger initial allocations and the latter would now more easily accept smaller initial allocations.

9.5 Conclusions

Quota tradability and joint implementation have been discussed in their potential roles of improving efficiency of a nontradable quota treaty. In doing so, we have seen that these two institutions may in fact help DCs or, more precisely, help IC signatories by helping DCs.

We saw in Section 9.3 that joint implementation (between an IC and a DC) as an approach to improving the efficiency of a nontradable quota treaty, has drawbacks, particularly in terms of serious problems of estimating net emission reductions from individual projects, and hence limiting the use of low-cost emission abatement in DCs. However, joint implementation was seen to have significant advantages in two respects: it is known to be politically acceptable, as shown by the FCCC, and is easy to start up.

In Section 9.4, we set out from the situation where ICs have committed themselves to emissions levels below their no control baseline, as is the case for signatories to an NTQ treaty, e.g., a more demanding version of the FCCC of 1992. Could DCs be attracted to a TQ arrangement with the IC signatories, so that it would benefit not only the ICs and possibly the emission reduction targets, but also DCs whose low-cost abatement options are so important to use in order to attain efficiency in an international carbon reduction treaty? DCs would, of course, not accept to join such a treaty unless they are given initial emission allocations that more than fully compensate them for doing so. If emission allocations are sought where DCs are compensated, but barely so, we have seen that these allocations must be based on an upper bound to the estimated *MACs* for DCs and a lower bound to the estimated *MACs* for ICs. More likely, at least for a hypothetical TQ treaty in the near future, initial allocations to DCs would approach a level represented by their estimated maximum baseline emissions. If so, we have seen that DCs would benefit more, the higher the differences in marginal abatement costs between ICs and DCs. Somewhat surprisingly, we also saw that, for the case where initial DC allocations are close to the upper bound, unexpected increases in the value of fossil fuel use in DCs would make their gain from joining a TQ treaty even larger.

Thus, making nontradable quotas tradable, e.g., those quotas that may eventually be the result of (more ambitious versions of) the FCCC, and attracting DCs to participate would allow reductions of carbon emissions to be conducted in an internationally efficient manner (barring incorrigible market imperfections in global quota trade). This efficiency gain could also increase the speed at which an even more ambitious international carbon reduction treaty could be arrived at. By contrast, we have seen that joint implementation may actually reduce the prospects for moving in that direction.

Notes

[1] For simplicity, we assume here that quotas as well as carbon taxes refer to the carbon content of fossil fuel and not to the emissions themselves. In a short-run perspective, this is probably not a severe constraint on the analysis. In the longer run, however, ways to remove carbon from emissions may become economically worthwhile, which require policy instruments to be designed taking this possibility into account. Then, to provide the appropriate incentives for R&D of new technologies, a treaty would, in principle, have to be such that proof of carbon recovery would be credited to firms and countries that employ such techniques.

[2] The Montreal Protocol allows trade of CFC *production* quotas generally and CFC *use* quotas among countries within the European Community, treated as one party to the Protocol. (I do not know whether such trade has taken place to any significant extent.) However, the most important kind of trade, that of CFC use quotas among all the parties to the Protocol, is not allowed.

Acknowledgments

Helpful comments by Scott Barrett, Ger Klaassen, Per Molander, and two anonymous referees on an earlier draft are gratefully acknowledged.

References

Barrett, S., 1992, "Acceptable" Allocations of Tradable Carbon Emission Entitlements in a Global Warming Treaty and Transfers and Gains from Trading Carbon Emission Entitlements in a Global Warming Treaty, in *Combating Global Warming*, Chapters VI and VII, UNCTAD, United Nations, New York, NY, USA.

Barrett, S., 1993a, *Joint Implementation for Achieving National Abatement Commitments in the Framework Convention on Climate Change*, Revised draft, London Business School (for the OECD), 19 April, London, UK.

Barrett, S., 1993b, *A Strategic Analysis of "Joint Implementation" Mechanisms in the Framework Convention on Climate Change*, Preliminary draft, London Business School (for UNCTAD), 26 May, London, UK.

Bohm, P., 1990, *Efficiency Issues and the Montreal Protocol on CFCs*, World Bank Environment Working Paper No. 40, Washington, DC, USA.

Bohm, P., 1992, Distributional Implications of Allowing International Trade in CO_2 Emission Quotas, *The World Economy* **15**(1):107–114.

Bohm, P., 1994, *The Feasibility of Joint Implementation of Carbon Emissions Reductions*, Paper presented at IPCC Working Group III Workshop, 17–20 January, Tsukuba, Japan.

Bohm, P., and Larsen, B., 1994, Fairness in a Tradable-Permit Treaty for Carbon Emissions Reductions in Europe and the Former Soviet Union, *Environmental & Resource Economics*, **4**(3):219–239.

Dudek, D., and LeBlanc, A., 1992, *Preserving Tropical Forests and Climate: The Role of Trees in Greenhouse Gas Emissions Trading*, Paper presented at the workshop on "Sustainable Development" (for UNCTAD), Rio de Janeiro, June 1–4, The Environmental Defense Fund, New York, NY, USA.

Grubb, M., 1989, *The Greenhouse Effect: Negotiating Targets*, Royal Institute of International Affairs, London, UK.

Hanisch, T., 1991, *Joint Implementation of Commitments to Curb Climate Change*, Policy Note 1991:2, CICERO, Oslo, Norway.

Hanisch, T., Pachauri, R., Schmitt, P., and Vellinga, P., 1992, *The Climate Convention: Criteria and Guidelines for Joint Implementation*, Policy Note 1992:2, CICERO, Oslo, Norway.

Hoel, M., 1993, Harmonization of Carbon Taxes in International Climate Agreements, *Environmental & Resource Economics* **3**:221–231.

Jones, T., 1993, *Operational Criteria for Joint Implementation*, OECD/IEA International Conference on the Economics of Climate Change, 14–16 June, Paris, France.

Nordhaus, W., 1991, The Cost of Slowing Climate Change: A Survey, *The Energy Journal* **12**(1):87–108.

Chapter 10

Control of Reciprocal Transboundary Pollution and Joint Implementation

Andries Nentjes
Department of Economics and Public Finance
University of Groningen, Groningen, The Netherlands

Abstract

International conventions to protect the environment usually have the form of commitments to reduce emission on a reciprocal basis. This chapter analyzes the conditions under which a Pareto-efficient agreement to reduce emissions reciprocally can be improved by joint implementation. Joint implementation is conceived as a second phase in which obligations to reduce emissions, which were agreed upon in the first phase, are exchanged for money. The scope for reciprocal reduction of emissions and for joint implementation are discussed first for a case where each government cares for its national environment only, and next for the case where a government is also willing to pay for improvement of the environment abroad.

Key words: joint implementation, international environmental conventions, reciprocal transboundary pollution.

10.1 Introduction

Despite the lip-service that is paid to the "polluter-pays" principle as a guideline for settling international environmental relations, the standing practice is that the state which is the victim of transboundary pollution has to pay the polluting country to reduce its emissions. Where environmental spillovers are unilateral the victim can pay a sum of money to compensate the polluter for the extra cost of pollution control. The solution could also be applied when pollution is reciprocal: parties could compensate each other reciprocally for their extra abatement efforts and the convention combines obligations to reduce emissions with payments in money. In their analyzes most economists have taken for granted that countries will proceed according to this scheme. Mäler (1989, 1990) and Welsch (1993), for example, take a noncooperative Nash solution as a reference point and compare it with a Pareto-optimal solution that is realized by allowing money payments between parties. However, this approach is more normative than positive. It does not reflect what really happens. International environmental conventions usually take the form of reciprocal promises to restrict emissions. In other words, countries pay each other with reductions of their national emissions instead of paying in money. This observation raises the question of whether a Pareto-optimal agreement in terms of reciprocal reductions of emissions will result in the same rates of emission reductions and will raise welfare in participating countries to the same degree as will a treaty that combines emission reductions with money payments between parties.

In his analysis of international cooperation to reduce emissions of a global pollutant, Hoel (1991) has shown that indeed there is a difference between a Pareto optimum attained by reciprocal reductions of emissions and the outcome of an effort to reduce emissions in such a way that joint welfare of participating countries is maximized. He does not work out the full implications of the distinction between the two arrangements. The main purpose of this chapter is to analyze under which conditions, and to what extent, the welfare gain from an international agreement to control emissions reciprocally that has been concluded and ratified can be improved by adding a second phase in which parties are allowed to recontract, by proposing an increase or decrease in their obligations to control pollution in exchange for money. The analysis concentrates on a nonglobal pollutant: the country where the pollutant is emitted does affect environmental damage in the countries that are involved in reciprocal pollution. The global pollutant can be viewed as a special case of this.

The relevance of our model is that it gives the political concept of joint implementation a proper place in economic theory. Although not explicitly mentioned, the idea of joint implementation is implicitly part of the UN Montreal Protocol of September 1987 on substances that deplete the ozone layer.[1] The concept also pops up in the UN Framework Convention on Climate Change (May 1992). In order to curb emissions of greenhouse gases, parties to the Convention may implement their control policies and measures jointly with other parties (Article 4.2a). The Convention also defines "a mechanism for the provision of financial resources on a grant or concessional basis" for the implementation of the convention (Articles 11.1 and 5).

As will be shown, joint implementation, conceived as a financial mechanism that supplements agreements on reciprocal reductions of emissions, improves both the cost-effectiveness and environmental effectiveness of the conventions. Cost-effectiveness of joint implementation implies that a chosen level of environmental quality in country A can be achieved at lower cost by decreasing emissions in country B and increasing them in country A. Cost savings in A enable the country to compensate country B for its extra abatement effort. Environmental effectiveness of joint implementation means that environmental quality in country A can be improved by additional reductions of emissions in country B. One can imagine a situation where the marginal costs of improving environmental quality in A exceed the perceived marginal benefits in A, whereas country B can do the job at marginal costs which are below A's marginal benefits. A's environmental benefits make it worthwhile to compensate B for its additional abatement costs.

A second issue to be discussed is the way in which the terms of international contracts to reduce emissions as well as joint implementation are affected if the government of one country cares for the quality of the environment in another country. For example, governments of West European countries might be willing to contribute to emission reductions in Eastern Europe because environmental improvements in those countries give benefits to the donor countries. We shall define this as a case of international environmental care.

This chapter has been inspired by observations of existing practices and recent innovations in international bargaining on environmental issues, although the analysis is theoretical. Simplifying assumptions have had to be made, and the model abstracts from a host of problems that have to be taken into account in real negotiations and joint implementation. The following offers an apparatus of thought; not a set of directly applicable instructions.

The chapter is organized as follows. Section 10.2 gives a survey of the essential concepts and analyzes the properties of international agreements based on reciprocal reductions of emissions. In Section 10.3 the scope for joint implementation in the case of reciprocal transboundary pollution is discussed. Section 10.4 focuses on how international care affects environmental agreements. Conclusions are presented in Section 10.5.

10.2 Reciprocal Reductions of Emissions

10.2.1 Introduction

The analysis takes as its starting point two countries. Environmental quality in each country is affected by its own emissions and (for at least one of the countries) also by the emissions of the other country. Decisions on environmental policy are made by the national government. Its perception and valuation of the environmental problem are caught in a simple separable utility function generally used (e.g., Mäler, 1990; Hoel, 1991) of the form $W_i = B_i(r_i) - C_i(z_i)$, with $i = 1{,}2$. For country 1, B_1 represents the benefits from reducing the pollution load (r_1) and C_1 the costs of emission reduction (z_1). The concept of environmental benefits does not necessarily imply that benefits can be measured accurately and "translated" into money terms. What is meant here is benefits in a subjective sense; it is assumed that the political body is willing to pay for environmental improvement. Environmental benefit and cost functions are twice differentiable with decreasing marginal benefits and increasing marginal costs. Reductions of emissions and pollution loads are connected through the transmission relations $r_1 = r_1(z_1, z_2)$ and $r_2 = r_2(z_1, z_2)$. The transmission functions are specified in a linear form: $r_1 = a_{11}z_1 + a_{12}z_2$ and $r_2 = a_{21}z_1 + a_{22}z_2$; where a_{11}, a_{12}, a_{21} and a_{22} are the transmission coefficients. This specification implies that we consider nonuniformly dispersed (nonglobal) pollutants, such as acidifying emissions.

10.2.2 Noncooperative Nash equilibrium

Each of the governments pursues the national interest as it sees it. In the economic approach such behavior is stylized by maximizing the government's utility function under the constraint of the relevant transmission equation. When there is no cooperation between governments, and each government considers the level of emissions and the emission reduction of its neighbor as given, the condition for the noncooperative Nash equilibrium is derived by maximizing the Lagrange functions:

$$L_1 = B_1(r_1) - C_1(z_1) + \lambda_1(r_1 - a_{11}z_1 - a_{12}z_2), \tag{10.1}$$

$$L_2 = B_2(r_2) - C_2(z_2) + \lambda_2(r_2 - a_{21}z_1 - a_{22}z_2). \tag{10.2}$$

The first-order conditions are:

$$C_1' = a_{11}B_1' \tag{10.3}$$

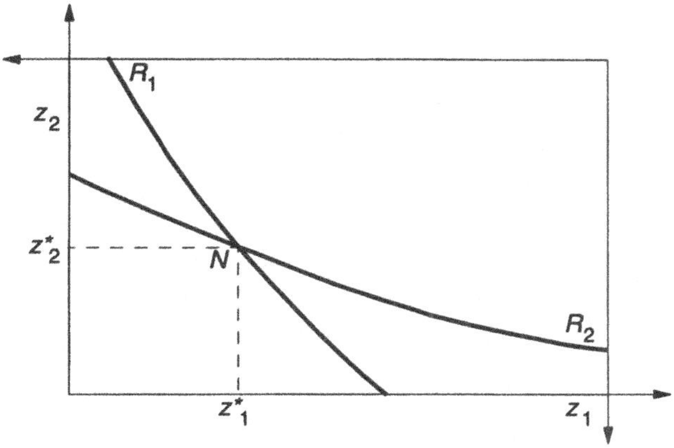

Figure 10.1. Noncooperative Nash equilibrium.

$$C_2' = a_{22}B_2' \tag{10.4}$$

Each government controls its emissions up to the point where marginal costs are equal to the marginal internal, or national environmental benefits. Equations (10.3) and (10.4) are both functions in z_1, z_2 the so-called reaction functions. By substituting the transmission functions in the first-order conditions, and taking the total differential, the sign and magnitude of the reaction coefficients can be found. For country 1 the reaction coefficient is:

$$\frac{\mathrm{d}z_1}{\mathrm{d}z_2} = \frac{-a_{12}}{a_{11} - (C_1''/a_{11}B_1'')} < 0, \quad > -1. \tag{10.5}$$

If country 2 intensifies its emission reduction effort, country 1 will react by reducing its abatement effort (and by increasing its emissions). The reaction curve of country 1 is represented by curve R_1 in *Figure 10.1*; R_2 is the reaction curve of country 2.[2] The noncooperative Nash equilibrium is presented in *Figure 10.1* by the point N, which is the intersection of the two reaction curves (10.3) and (10.4).

10.2.3 Pareto efficiency

Both countries can improve their welfare by coordinating their pollution control policies. Country 1 can offer to raise its emission reduction effort on the condition that country 2 follows the same line of action. The effect will be that country 1 gets a greater reduction of pollution load in return for an extra unit of emission reduction than it would have had under the

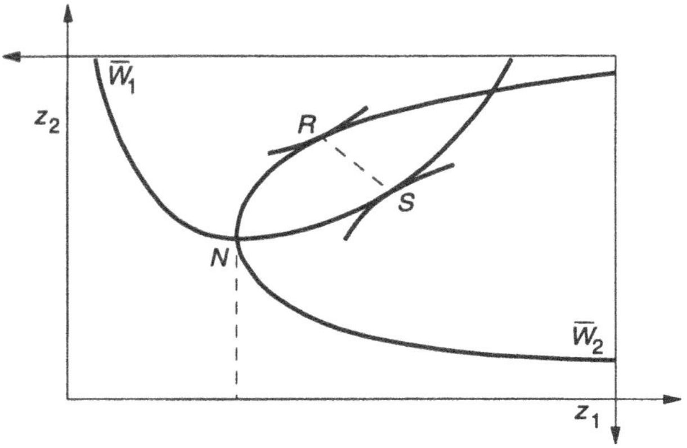

Figure 10.2. Cooperative Nash equilibrium.

noncooperative policy regime. The same holds true for country 2. There is room to contract a reciprocal increase in emission reductions from which both countries will derive net benefits. This is shown in *Figure 10.2.* $\overline{W}_1$ is the iso-welfare curve or iso-net benefit curve of country 1.[3] Any point in the area above $\overline{W}_1$ constitutes higher welfare for country 1 than points on the iso-welfare curve and therefore is preferred to points on $\overline{W}_1$. *Figure 10.2* also shows the iso-welfare curve for country 2, $\overline{W}_2$. Points to the right of $\overline{W}_2$ are preferred to points on the curve. The contract area, enclosed by curves $\overline{W}_1$ and $\overline{W}_2$ is the set of all possible combinations of z_1, z_2 which are Pareto-superior to N.

The first-order condition for a Pareto-optimum is derived by maximizing the Lagrange function:

$$\begin{aligned} L \;=\; & B_1(r_1) - C_1(z_1) + \lambda_1(r_1 - a_{11}z_1 - a_{12}z_2) \\ & + \mu_1[(\overline{W}_2 - B_2(r_2) + C_2(z_2)] + \lambda_2(r_2 - a_{21}z_1 - a_{22}z_2). \end{aligned} \tag{10.6}$$

The first-order conditions are:

$$C_1' = a_{11}B_1' - \mu_1 a_{21}B_2', \tag{10.7}$$

$$C_2' = a_{22}B_2' - 1/\mu_1 a_{12}B_1'. \tag{10.8}$$

Equations (10.7) and (10.8) state that for a Pareto optimum the transboundary benefits of pollution control have to be taken into account alongside the national or internal benefits. The Lagrange multiplicator is $-\mu_1 > 0$; this implies that in the Pareto optimum C_1', C_2' and consequently z_1, z_2 will be larger than in the noncooperative Nash equilibrium.[4]

The first-order conditions can be reduced to one equation:

$$\frac{C_1' - a_{11}B_1'}{a_{12}B_1'} = \frac{a_{21}B_2'}{C_2' - a_{22}B_2'}, \tag{10.9}$$

which is one equation with two variables z_1, z_2. This reminds us that the Pareto optimum consists of a range of combinations z_1, z_2. If negotiations concentrate on reciprocal emission reductions and no money payments are involved, the range of Pareto-efficient solutions is represented by the contract curve *RS* in *Figure 10.2*, bounded by the restrictions $W_1 \geq \overline{W}_1, W_2 \geq \overline{W}_2$; where $\overline{W}$ is the level of net benefits in the noncooperative Nash equilibrium.

From equation (10.9) it follows that

$$a_{12}B_1'dz_2 = -(C_1' - a_{11}B_1')dz_1, \tag{10.9a}$$

$$a_{21}B_2'dz_1 = -(C_2' - a_{22}B_2')dz_2, \tag{10.9b}$$

The right-hand side of (10.9a) represents the marginal net cost of increasing z_1 for country 1; the left-hand side states the marginal benefits country 1 gets in return through the increase in z_2 by country 2. In short, (10.9a) and (10.9b) state that in the Pareto optimum of reciprocal reductions, the marginal costs equal the marginal benefits of the bargain for each of the two countries.

However, from the first-order conditions (10.7) and (10.8) it also follows that in the Pareto optimum of reciprocal emission reductions the following set of equalities is possible:

$$a_{21}B_2' > C_1' - a_{11}B_1' \tag{10.7a}$$

$$C_2' - a_{22}B_2' > a_{12}B_1' \tag{10.8a}$$

In words, it could be that the marginal transboundary benefits for country 2 arising from pollution control in country 1 exceed the marginal net costs of emission reductions in country 1; simultaneously, the marginal benefits in country 1 of z_2 are lower than the net marginal cost of z_2 for country 2. The situation is illustrated in *Figure 10.3* by the pair ${}^{M}z_1, {}^{M}z_2$ and suggests that further improvements in welfare in both countries are possible by increasing z_1 and decreasing z_2. However, once the Pareto optimum of reciprocal emission reductions has been realized, such a reshuffling is not possible since it would be in conflict with the conditions (10.9a) and (10.9b). For example, increasing z_1 would increase the net costs of country 1 and reducing z_2 would decrease its benefits.

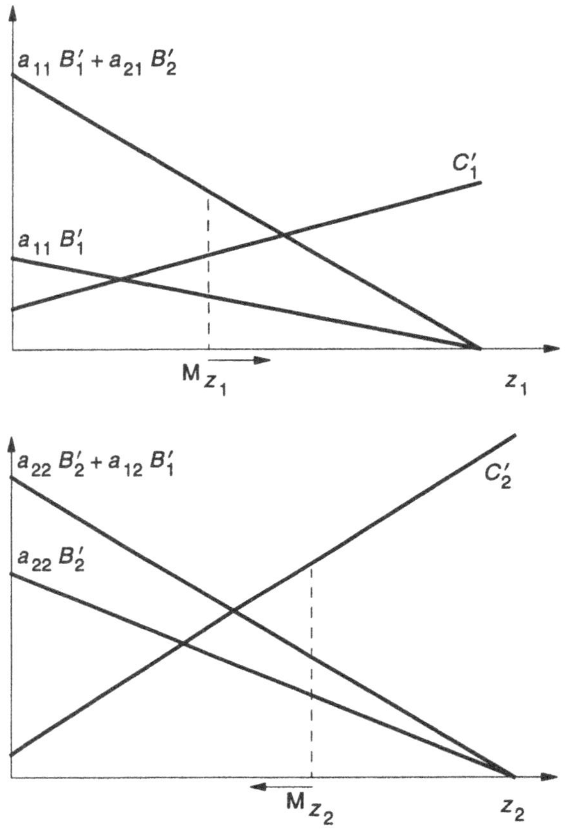

Figure 10.3. From cooperative Nash equilibrium to joint maximum benefit.

10.2.4 Market equilibrium

Where countries 1 and 2 negotiate on the size of their emission reductions the ratio z_1/z_2 defines the quantity of emission reductions country 1 can get in return from country 2 when it reduces its emissions by 1 unit: the ratio z_2/z_1 constitutes the rate of exchange or price of the "goods" (emission reductions) that are traded.

A possible procedure for finding an equilibrium solution within the range of Pareto-efficient solutions, that is, a point on the contract curve, would be the introduction of a market for the two goods. A perfectly competitive market could be simulated if the negotiating parties accept as a third party an auctioneer who announces price ratios and if each negotiator would take the price ratio as given and reveal the true amount of emissions control it is willing to offer at the price proposed by the auctioneer. Under these conditions it is possible to derive an offer curve for each of the two negotiators and a market-clearing equilibrium at the point where the two offer curves intersect.

The behavior of each negotiator under the market regime is modeled by maximization of its utility function under the constraint of a given price ratio. The Lagrange functions are:

$$L_1 = B_1(r_1) - C_1(z_1) + \lambda_1(r_1 - a_{11}z_1 - a_{12}\pi_1 z_1), \tag{10.10}$$

$$L_2 = B_2(r_2) - C_2(z_2) + \lambda_2(r_2 - a_{21}\pi_2 z_2 - a_{22}z_2), \tag{10.11}$$

where π_1 is the price ratio z_2/z_1 and $\pi_2(= 1/\pi_1)$ the price ratio z_1/z_2. The constraint $r_1 - (a_{11} + a_{12}\pi_1)z_1$ implies that in deciding on its emission reduction, country 1 accounts for the direct impact of that reduction on the pollution load at home (that is, $a_{11}z_1$) and also for the indirect impact, since country 2 responds by offering z_2 at a given ratio π_1, which reduces the pollution load in country 1 by $a_{12}\pi_1 z_1 (= a_{12}z_2)$. The same type of argument holds for country 2. The first-order conditions are:

$$C_1' = (a_{11} + a_{12}\pi_1)B_1', \tag{10.12}$$

$$C_2' = (a_{22} + a_{21}\pi_2)B_2'. \tag{10.13}$$

Equations (10.12) and (10.13) are functions in z_1, z_2. They can be interpreted as offer curves. *Figure 10.4* gives a graphical derivation of the offer curve for country 1 as a series of points of tangency of successive price lines and iso-welfare curves. In the same way, the offer curve for country 2 can be derived. The two offer curves and their point of intersection, that is the market equilibrium M, are shown in *Figure 10.5*.[5] Within the range of possible price ratios an equilibrium rate of exchange π_1^e exists. Mathematically, the price ratio that follows from (10.12) and (10.13) is:

$$\frac{z_2}{z_1} = \pi_1 = \frac{C_1' - a_{11}B_1'}{a_{12}B_1'} = \frac{1}{\pi_2} = \frac{a_{21}B_2'}{C_2' - a_{22}B_2'}. \tag{10.14}$$

A comparison with (10.9) reveals that the market equilibrium satisfies the conditions for Pareto efficiency in the case of reciprocal reduction of emissions. Equation (10.14) also shows that a finite price > 0 is possible only if transboundary pollution is reciprocal (a_{21} and $a_{12} > 0$), if there are (politically perceived) positive marginal environmental benefits in both countries, and abatement technologies are available at finite marginal costs.

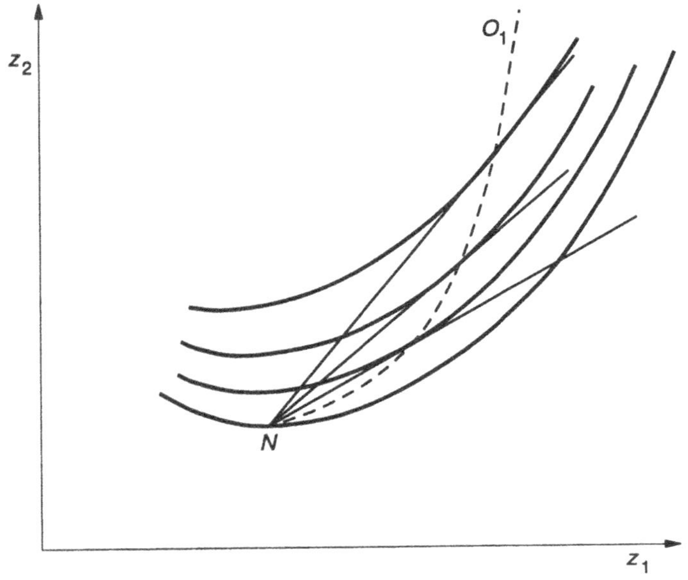

Figure 10.4. Offer curve of country 1.

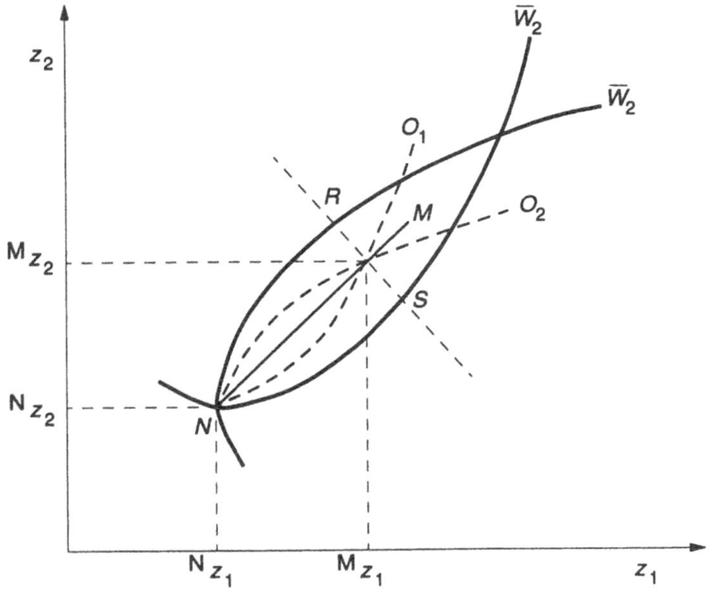

Figure 10.5. Market equilibrium of reciprocal reduction.

Reciprocal transboundary pollution may be very asymmetric, but this does not prevent a deal with a finite rate of exchange being struck.

It should be remembered that the market solution as a particular case of the range of cooperative Nash solutions is possible only if the negotiating parties accept the rules of the game by adjusting their emission reductions offers to the prices mentioned by the auctioneer and do not behave strategically. There is strategic behavior if negotiators try to influence the market price through their offers of emission reductions. For example, country 1 could understate its willingness to control emissions at given prices if it expects that this will raise the market equilibrium price in its favor. The result would be that the market equilibrium is not a point on the curve of Pareto-efficient solutions. The assumption that parties do not behave strategically limits the applicability of our "market model" to real-world cases. The usefulness of the concept of market equilibrium is that it fixes a point on the contract curve and provides a focus for the outcome of negotiations on reciprocal reductions of emissions. In the next section the market equilibrium concept will function as a point of reference that is needed for an analysis of the scope for joint implementation.

10.3 Joint Implementation

10.3.1 Introduction

The Pareto efficiency condition (10.9) is derived for a situation where the welfare of two countries can only be increased by exchanging reductions of emissions. In this section it will be demonstrated that Pareto-efficient outcomes with larger welfare gains for both countries become feasible by introducing an extra means of exchange that has utility for both parties. Let the means of exchange be international money. The variables Y_1, Y_2 are inserted in the welfare functions; where Y_1 is the revenue received by country 1, and Y_2 is the revenue of country 2. Joint revenue is constrained in the sense that revenue for country 1 is expenditure for country 2, and vice versa; i.e., $+Y_1 = -Y_2$. A welfare function in the form $W_i = B_i - C_i \pm Y_i$ implies that all these components are expressed in international money units. The use of international money makes it feasible to compensate a country that increases its pollution control, which will decrease its net environmental benefits, by increasing its real income. In the other country real income is reduced, but this can be compensated by higher transboundary environmental benefits and lower net costs of pollution control, brought about by lowering its abatement.

10.3.2 Pareto efficiency

The first-order conditions for Pareto efficiency with payments are derived by maximizing the Lagrange function:

$$\begin{aligned} L &= B_1(r_1) - C_1(z_1) + Y_1 + \mu_1 \left\{ \overline{W}_2 - B_2(r_2) + C_2(z_2) - Y_2 \right\} \\ &\quad + \lambda_1(r_1 - a_{11}z_1 - a_{12}z_2) + \lambda_2(r_2 - a_{21}z_1 - a_{22}z_2) + \\ &\quad \delta(Y_1 + Y_2). \end{aligned} \tag{10.15}$$

The first-order conditions are:

$$C_1' = a_{11}B_1' + a_{21}B_2', \quad \text{or} \quad C_1' - a_{11}B_1' = a_{21}B_2' \tag{10.16}$$

$$C_2' = a_{21}B_1' + a_{22}B_2', \quad \text{or} \quad C_2' - a_{22}B_2' = a_{12}B_1' \tag{10.17}$$

The two unknown variables z_1, z_2 can be solved from equations (10.16) and (10.17); these equations are special cases of (10.7) and (10.8), with $-\mu_1 = 1$. In other words, a Pareto-efficient solution of emission reductions with money transfers must be a point on the Pareto efficiency curve with emission reductions as the only active variables. Therefore, finding Pareto-efficient solutions with money payments can be interpreted as a movement along the z_1, z_2 Pareto efficiency curve, which is the dotted curve through R and S in *Figures 10.2* and *10.5.* The welfare of each country is interpreted as the sum of its net willingness to pay for a clean environment plus real income. Starting from a point on the curve RS, for example, M (*Figure 10.5*), it is possible to assess whether a movement along the curve, which changes the combination (z_1, z_2), increases the net environmental benefits of, say, government 1 by more or by less than it decreases welfare of country 2. If the country whose net environmental benefits increase can and does compensate fully the country whose net environmental benefits decrease by means of a side payment in money, both parties can agree on the move along the contract curve. In this way the Pareto optimum with money transfers can be found. It should be noted that since international money is involved it cannot be excluded that the Pareto optimum is a point beyond the disagreement point R or S on the curve through R and S in *Figure 10.5.* This is so since a country whose welfare is in danger of being reduced below the welfare of the Nash noncooperative equilibrium because of its increased costs of pollution control, can be compensated in money.

An alternative way to clarify the impact of money payments on emission reduction is shown in *Figure 10.3.* As the market equilibrium of reciprocal emission reductions without money payments, we have ${}^{\mathrm{M}}z_1, {}^{\mathrm{M}}z_2$, with

$B_2'({}^{M}z_1, {}^{M}z_2) > C_1'({}^{M}z_1) - B_1'({}^{M}z_1, {}^{M}z_2)$. This implies that country 2 is willing to pay country 1 in money for increasing its level of abatement. Simultaneously, country 2 could reduce its own level of emission control, save abatement costs and set money free for compensating country 1 for both its net marginal cost of increasing z_1 and for its loss of transboundary marginal benefits arising from country 2's decrease in pollution control. This is possible since $C_2'({}^{M}z_2) - B_2'({}^{M}z_1, {}^{M}z_2) > B_1'({}^{M}z_1, {}^{M}z_2)$. The Pareto optimum with money payments is attained at z_1^*, z_2^*, where the two inequalities change into equalities.[6]

In the real world agreements on reductions of transboundary pollution usually fix the size of emission reductions for each party to the agreement. The above analysis makes it clear that, even if a Pareto-optimal solution were to be realized, such a best of all possible contracts in terms of reciprocal reductions leaves scope for additional gains in net benefits for both parties by allowing recontracting, taking the commitments to control emissions reciprocally as a starting point. In this second phase obligations to reduce emissions are exchanged for money. Joint implementation clauses in recent international conventions on climate change and reductions of CFCs indicate that policy makers are beginning to discover the limitations of reciprocal reductions and the possibility of welfare gains by reallocating pollution control among countries in exchange for payment in money.

From (10.14) it follows that in a special case where pollution control cost and benefit functions in the two countries are identical and their internal pollution coefficients a_{11}, a_{22}, as well as their pollution import coefficients a_{12}, a_{21}, are equal then, the exchange ratio $z_2/z_1 = 1$ in the case of reciprocal reductions of emissions. Both countries reduce their emissions by the same amount. From (10.16) and (10.17) it appears that in this case of symmetrical transboundary pollution between countries with identical welfare functions, there is no scope for joint implementation since the joint maximum welfare can be attained by reciprocal reductions of emissions. This result illustrates that in the case of Pareto-optimal contracts of reciprocal emission reductions, joint implementation can be viewed as an instrument that corrects for differences in cost and benefit functions together with asymmetries in transboundary pollution.

10.3.3 Market equilibrium

The preceding analysis has demonstrated that there exists a Pareto optimum with money payments, although it does not say how it can be found. One possible solution is to simulate a perfectly competitive market by introducing an auctioneer who announces prices. The difference from the market solution discussed in Section 10.2.4 is that the use of money makes it feasible to split up the negotiations on z_1, z_2 into separate decisions on z_1 and z_2. The auctioneer operates on and coordinates two "markets": for z_1 the price is t_1, and for z_2 the price is t_2. The rule of the game is that at the announced price t_1, country 1 reports how many units of z_1 it is willing to offer, or to withdraw from the market compared with the initial situation; country 2 states how many units z_1 it demands, because of the transboundary environmental benefits they create. The demand of country 2 can be positive or negative. *Mutatis mutandis*, the supply of and demand for z_2 are discovered in the same way.

Formally, the problem is solved as follows. Let T_1 be the money payment received by country 1 with $T_1 = t_1 z_1$, where t_1 is the compensation in money received per unit of emissions reduced. $T_2 = t_2 z_2$ is the sum paid by country 1 and received by country 2. Both countries maximize their welfare, including net transfers. Maximization of the Lagrange equations

$$L_1 = B_1(r_1) - C_1(z_1) + t_1 z_1 - t_2 z_2 + \lambda_1(r_1 - a_{11} z_1 - a_{12} z_2), \quad (10.18)$$

$$L_2 = B_2(r_2) - C_2(z_2) + t_1 z_1 - t_2 z_2 + \lambda_2(r_1 - a_{21} z_1 - a_{22} z_2), \quad (10.19)$$

gives the first-order conditions:

$$a_{11} B_1' - C_1' = t_1, \quad (10.20)$$

$$a_{21} B_2' = t_1, \quad (10.21)$$

$$a_{22} B_2' - C_2' = t_2, \quad (10.22)$$

$$a_{12} B_1' = t_2. \quad (10.23)$$

Equation (10.20) defines the supply of z_1 by country 1, and (10.21) the demand of country 2 for z_1; (10.22) and (10.23) are the supply of and demand for z_2. From the four equations that define the market equilibrium, the four unknown values t_1, t_2, z_1 and z_2 can be solved. In other words, there exists a price vector t_1, t_2 where the demand for z_1, z_2 equals the supply, The conditions for a market equilibrium with money payments also satisfy the condition for a Pareto optimum with money payments. This can be verified by substituting (10.21) in (10.20) and (10.23) in (10.22).

The reader should keep in mind that the analysis rests on the same behavioral assumptions as the analysis of Section 10.2.4: parties accept an auctioneer and reveal their true preferences. An alternative adjustment mechanism that claims not to be sensitive to strategic behavior is given by Chander and Tulkens (1990, 1992).

10.3.4 Cost-effective joint implementation

One can imagine that countries that consider joint implementation have an objective in mind that is less ambitious than maximizing net benefits. Instead it could be cost-effectiveness: negotiators set out to reallocate the initial result from reciprocal reduction $({}^{M}z_1, {}^{M}z_2)$ in such a way that the joint costs of pollution control are minimized under the constraint that the pollution loads $({}^{M}r_1, {}^{M}r_2)$ that result from the agreement on reciprocal reductions of emissions do not increase. Klaassen *et al.* (1994) and Førsund and Naevdal (Chapter 11) discuss procedures for finding the cost minimum for n countries by way of successive bilateral negotiations. In this section we only wish to assess whether a Pareto optimum that (by assumption) has emerged from negotiations on reciprocal reductions of emissions leaves scope for joint implementation geared to cost-effectiveness.

The cost minimum is found by maximizing the function

$$\begin{aligned} L &= -C_1(z_1) - C_2(z_2) + \lambda_1(\bar{r}_1 - a_{11}z_1 - a_{12}z_2) \\ &\quad +\lambda_2(\bar{r}_2 - a_{21}z_1 - a_{22}z_2), \end{aligned} \tag{10.24}$$

where r_1, r_2 are the concentrations of pollutants that follow from the emission reductions that have been agreed. It should be noted that the pollution constraints have the form of inequalities. The first-order condition is:

$$\frac{C_1'}{C_2'} = \frac{-\lambda_1 a_{11} - \lambda_2 a_{21}}{-\lambda_1 a_{21} - \lambda_2 a_{22}}, \tag{10.25}$$

where $-\lambda_1, -\lambda_2$ are the shadow prices of reducing the pollution loads in countries 1 and 2, respectively.[7]

Neither the Pareto efficiency condition (10.9), nor the market solution for reciprocal emission reductions (10.14), do meet the condition for cost-efficiency. This implies that even if negotiations on reciprocal reductions of emissions have led to a Pareto-efficient agreement there is still room left for the reallocation of obligations to reduce emissions between countries in order to realize the same environmental benefits against lower cost. Again this would ask for joint implementation. The country with relatively low marginal abatement costs (say country 1) can take over the obligation to reduce emissions from country 2, which has relatively high marginal abatement costs, and can receive compensation for its additional costs from the country that can increase its emissions. This case is illustrated in *Figure 10.6.* The total costs of pollution control are lowered (represented by a downward shift of the iso-cost curve) by substituting the initial uniform

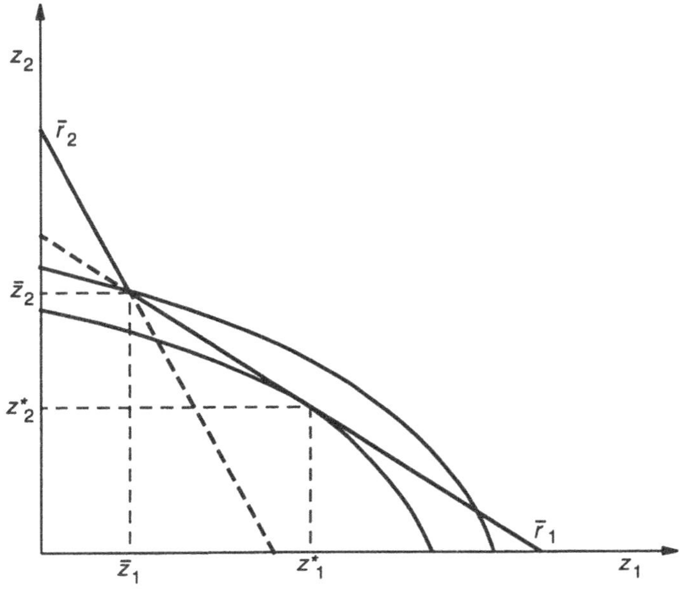

Figure 10.6. Cost-effective joint implementation.

distribution $\bar{z}_1, \bar{z}_2$ by z_1^*, z_2^*. The binding receptor $\bar{r}_1$ defines the trade ratio; $\mathrm{d}z_2/\mathrm{d}z_1 = -a_{11}/a_{12}$. In the example, the pollution load in country 1 and consequently its total environmental benefits remain constant. In country 2 the pollution load decreases and total environmental benefits increase. In this specific case country 2 has a double motive to pay country 1: lower costs as well as higher benefits.

10.4 International Environmental Care

10.4.1 Introduction

In this section we stick to the assumption that transboundary pollution is reciprocal, but drop the assumption that all countries are interested only in their own national environment and cost levels. It will be assumed that some countries are concerned about environmental quality in other countries. A major reason why the inhabitants of one country might care about environmental quality abroad could be their "selfish" interest in the conservation of nature or landscapes in other countries, which they can enjoy as tourists or by way of the media. A second reason could be genuine altruistic care for the quality of life abroad. The existence of such preferences could mean that the government of country 2 cares about the pollution load (r_1) in country 1.

10.4.2 Reciprocal reductions of emissions

If country 2 does care for the environment in country 1 then the environmental extra-territorial benefits of pollution control have to be added to the environmental benefits realized in country 2 itself. The welfare function of country 2 then takes the form:

$$W_2 = B_2(r_2) + {}^1B_2(r_1) - C_2(z_2), \tag{10.26}$$

where ${}^1B_2(r_1)$ represents the extra-territorial environmental benefits, or international environmental care. The introduction of international environmental care changes the noncooperative solution. If country 2 considers pollution control by country 1 as given, then the maximization of

$$\begin{aligned} L_2 &= B_2(r_2) + {}^1B_2(r_1) - C_2(z_2) + \lambda_1(r_1 - a_{11}z_1 - a_{12}z_2) \\ &\quad + \lambda_2(r_2 - a_{12}z_1 - a_{22}z_2) \end{aligned} \tag{10.27}$$

gives as the first-order condition:

$$C_2' = a_{22}B_2' + a_{12}{}^1B_2'. \tag{10.28}$$

A comparison with (10.4) reveals that for any given level of z_1 marginal cost C_2' and consequently z_2 will be higher than they would be without environmental care. This implies that the inclusion of altruistic preferences makes the reaction curve R_2 shift upward. Since country 1 maximizes its own welfare function with national benefits, only its reaction curve will not change position. Consequently, the Nash noncooperative solution will shift with z_2 higher and z_1 lower than in the case without environmental care (see also Hoel, 1991).

The existence of international environmental care also affects the cooperative solution. If countries 1 and 2 were to negotiate on reciprocal reductions of emissions and to behave as price takers, then the offer curve for country 2 can be derived from maximizing

$$\begin{aligned} L_2 &= B_2(r_2) + {}^1B_2(r_1) - C_2(z_2) + \lambda_1(r_1 - a_{11}\pi_2 r_2 - a_{12}z_2) \\ &\quad + \lambda_2(r_2 - a_{21}\pi_2 z_2 - a_{22}z_2), \end{aligned} \tag{10.29}$$

from which we have

$$C_2' = (a_{22} + a_{21}\pi_2)B_2' + (a_{11}\pi_2 + a_{12}){}^1B_2'. \tag{10.30}$$

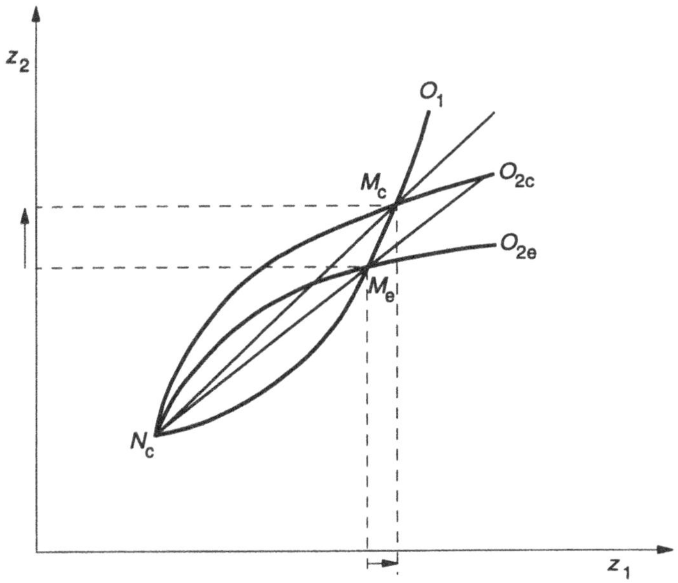

Figure 10.7. Market equilibrium with environmental care.

From (10.30) it follows that

$$\pi_2 = \frac{z_1}{z_2} = \frac{C_2' - a_{22}B_2' - a_{12}{}^1B_2'}{a_{21}B_2' + a_{11}{}^1B_2'}. \tag{10.31}$$

The properties that $-a_{12}{}^1B_2' < 0$ and $a_{11}{}^1B_2' > 0$ imply that for a given price π_2 the term $(C_2' - a_{22}B_2')$ must be larger, and/or that $A_{21}B_2'$ must be smaller than they are in equation (10.14). Therefore, the supply of z_2 by country 2 and its demand for z_1 are higher than they would have been without international environmental preferences. This implies a shift in country 2's offer curve from O_{2e} to O_{2c}, as indicated in *Figure 10.7.* International environmental care means that at each given level of z_1, country 2 is willing to offer a higher level of pollution control than it would have done if extra-territorial environmental benefits did not exist.

For country 1 the offer curve is specified by the equation

$$C_1' = (a_{11} + a_{12}\pi_1)B_1' \tag{10.12}$$

From (10.30) and (10.12) it follows that

$$\frac{z_2}{z_1} = \pi_1 = \frac{C_1' - a_{11}B_1'}{a_{12}B_1'} = \frac{1}{\pi_2} = \frac{a_{21}B_2' + a_{11}{}^1B_2'}{C_2' - (a_{22}B_2' + a_{12}{}^1B_2')}\ . \tag{10.32}$$

Country 2's greater willingness to control pollution (compared with preferences for the national environment only) means that in the market equilibrium the exchange ratio changes to the advantage of country 1, as *Figure 10.7* shows. It may be a surprise that the existence of environmental care raises the abatement effort of the donor country. The background is that to increase emission reductions at home is also a method to improve environmental quality in the neighboring country.

The general conclusion is that preferences for the environment abroad shift the burden of pollution control to the country with international environmental care. In the noncooperative Nash equilibrium its pollution control will be higher and, starting from that position, its willingness to offer pollution control at given exchange rates is higher too. The other country that is the object of environmental care will start from a lower noncooperative level of pollution control but this is compensated partly or perhaps even more than that by its higher level of abatement in the market solution induced by the more favorable rate of exchange.

10.4.3 Joint implementation

We are particularly interested in the improvements that could be brought about by joint implementation. Similar to the case with only preferences for the environment at home, there is no guarantee that with international environmental care the market solution of reciprocal reductions coincides with the Pareto-efficient outcome if money payments are used. The first-order conditions for a Pareto optimum with money payments are:

$$a_{11}(B_1' + {}^1B_2') + a_{21}B_2' = C_1', \tag{10.33}$$

$$a_{12}(B_1' + {}^1B_2') + a_{22}B_2' = C_2'. \tag{10.34}$$

Compared with the case of purely national preferences [equations (10.16) and (10.17)], the term ${}^1B_2'$ is added at the benefit side of the first-order conditions. This implies that both z_1 and z_2 are higher than they would have been without international environmental care. Usually $a_{11} > a_{12}$; *ceteris paribus* the additional pollution control, compared with (10.17) and (10.18), is mainly realized by a higher z_1 and not so much by higher z_2.

If country 2 is a rich country with high (subjective) national environmental marginal benefits there is a fair chance that in the market equilibrium of reciprocal emission reductions, its marginal costs of pollution control exceed the joint environmental benefits (represented by ${}^M z_2$ in *Figure 10.3*).

Care for the environment in the neighboring country raises z_2 even further. Under these conditions it is to be expected that under the joint implementation scheme the caring country 2 will pay country 1 to increase its pollution control and to accept a decrease in pollution control in country 2.

10.5 Conclusion

In the case of reciprocal transboundary pollution the possibilities to reduce emissions on a reciprocal basis can be constrained by differences in the costs and benefits of pollution control between countries, together with asymmetries in transboundary pollution. If the "purchase" of additional reductions of emissions abroad, by offering extra pollution control at home, is too expensive, because of the high net costs of pollution control, a country may still have the option to offer money in exchange for emission reductions to countries with relatively low marginal costs of pollution control and low internal environmental marginal benefits.

The ultimate cooperative equilibrium position can be realized by way of a procedure that consists of two steps. In the first step the negotiations have the form of reciprocal reductions of emissions, and the best possible (Pareto-optimal) solution that is feasible within the constraints of this procedure can be sought. The next step consists of reallocation of pollution control which is such that some parties increase their abatement of emissions and are paid for doing so by the countries which decrease their abatement effort.

This procedure allows us to give the political concept of joint implementation a place in economic theory. Negotiations usually concentrate on reciprocal reductions of emissions. Only recently have countries discovered the advantages of the second step, that is, revising the emission allocations of the first step and allowing payment in money to bring about the reallocation of obligations for control pollution. This is what joint implementation is in essence. In the real world, joint implementation might be a device for correcting inefficiencies in the initial allocation of reciprocal emission reductions, for example, when reduction is equiproportional. In this chapter it was shown that even if the initial allocation was Pareto-optimal, there is generally scope for further improvement in the joint implementation stage.

The conclusions do not change if there exists international environmental care. The caring country can encourage additional reductions of emissions in the country for whose environment it cares, by offering that country a more favorable rate of exchange in the negotiations on reciprocal reductions of emissions. As a complement, the country that cares can improve the efficiency of reciprocal reductions in the joint implementation stage by proposing payments for additional reductions of emissions in the country it cares for, and simultaneously increases emissions at home.

The analysis of international care leads to the conclusion that its existence does not necessarily demand a special kind of subsidy. The desired outcomes can be realized in the process of negotiation on reciprocal reductions of emissions. Joint implementation is an instrument for making the agreement more Pareto-efficient, just as it is when environmental care does not exist. This implies that in the simple case discussed in this chapter there is no need to take account of environmental care by way of special financial arrangements different from the joint implementation mechanism that functions between countries that care only for their own national environments.

Acknowledgments

I thank Marcus Amann, Janusz Cofalla, Paul Koutstaal, Tom Schelling, the editors, and two anonymous referees for their valuable comments and suggestions on earlier versions.

Notes

[1] According to Article 2.5 countries which are relatively small producers of CFCs are allowed to transfer to or receive from any other party to the Protocol production in excess of its limits, provided that the combined levels or production of the parties concerned do not exceed the production limit. Article 2.8 allows consumer countries which are members of a recognized regional economic integration unit to fulfill their obligations jointly with respect to consumption, provided their total combined consumption remains below the required limits.

[2] The negative reaction will be stronger, the higher the pollution import coefficient (a_{12} for country 1) and the sharper the increase of marginal abatement costs relative to the decline in marginal benefits.

[3] To the right of point N the $\overline{W}_1$ curve is increasing: any increase in z_1 increases pollution control costs; in order to raise environmental benefits in such a way that net benefits in country 1 remain constant, country 2 would have to increase its level of pollution control as well. Increasing marginal costs in country 1 mean that country 2 would have to raise z_2 increasingly in order to provide the necessary additional environmental benefits. To the left of point N, net environmental benefits of country 1 are reduced by decreasing z_1 because the loss of environmental benefits exceeds the savings on pollution control costs. In order to keep $\overline{W}_1$ constant country 2 has to increase its level of pollution control.

[4] The concepts of Nash noncooperative equilibrium and Pareto equilibrium are by now part of common wisdom in the analysis of international pollution problems; see Pethig (1982); Nentjes and Wiersma (1984); Nentjes (1990); Hoel (1991).

[5] The model of a market equilibrium for n countries is given in Nentjes (1990).

[6] The aggregate marginal environmental benefits of z_1 increase because z_2^* is lower than ${}^{M}z_2$; aggregate marginal benefits of z_2 decrease because z_1^* is higher than ${}^{M}z_1$.

[7] The constraint qualification asks that the number of binding constraints is less than the number of variables, therefore only one constraint can be binding in the cost minimum of our example.

References

Chander, P., and Tulkens, H., 1990, *Strategically Stable Costsharing in an Economic-Ecological Process*, Paper presented at congress EAERE, Stockholm, Sweden.

Chander, P., and Tulkens, H., 1992, Theoretical Foundations of Negotiations and Cost Sharing in Transfrontier Pollution Problems, *European Economic Review* **36**:388–398.

Hoel, M., 1991, Global Environmental Problems: The Effects of Unilateral Actions Taken by One Country, *Journal of Environmental Economics and Management* **20**:55–70.

Mäler, K.G., 1989, The Acid Rain Game, in H. Folmer and E. van Ierland, eds., *Valuation Methods and Policy Making in Environmental Economics*, Elsevier, Amsterdam, The Netherlands.

Mäler, K.G., 1990, International Environmental Problems, *Oxford Review of Economic Policy* **6**:80–107.

Klaassen, G., Førsund, F., and Amman, M., 1994, Emissions Trading in Europe with an Exchange Rate, *Environmental and Resource Economics*, forthcoming.

Nentjes, A., 1990, An Economic Model of Transfrontier Pollution Abatement, in V. Tanzi, ed., *Public Finance, Trade and Development*, Wayne State University Press, Detroit, MI, USA.

Nentjes, A., and Wiersma, D., 1984, Pareto-optimaal milieubeleid voor grensoverschrijdende vervuiling (*Pareto-optimal Environmental Policy for Transboundary Pollution*), Research Memorandum No. 169, Institute for Economic Research, University of Groningen, The Netherlands.

Pethig, R., 1982, Reciprocal Transfrontier Pollution, in H. Siebert, ed., *Global Environmental Resources: The Ozone Problem*, Peter Lang Verlag, Frankfurt am Main/Bern.

Welsch, H., 1993, An Equilibrium Framework for Global Pollution Problems, *Journal of Environmental Economics and Management* **25**:64–79.

Chapter 11

Trading Sulfur Emissions in Europe

Finn R. Førsund
Department of Economics, University of Oslo, Norway
Eric Nævdal
Foundation for Research in Economics and Business Administration
University of Oslo, Norway

Abstract

A second protocol on the reduction of sulfurous emissions is currently being negotiated by the European countries. Due to a number of factors the final protocol may not be the best of all possible agreements. This chapter suggests that a given protocol may be improved by allowing countries to trade the emissions they are allowed under the protocol, and suggests various institutional arrangements under which this may be done. The results of a computerized simulation of such trading are presented and used to evaluate the conditions under which such a trading scheme may be beneficial.

Key words: emissions trading, cost efficiency, environmental agreements.

11.1 Background

The first generation of European transboundary pollution agreements were based on uniform reductions of emissions as a basic principle of fairness. Thus the first sulfur protocol between European countries called for all signatories to cut back their sulfur emissions by 30% by 1993 compared with emissions in 1980. However, it is well known that uniform reductions in general imply higher total costs than necessary to reach the same environmental objectives, i.e., this principle of fairness is not cost-effective. The parties to the negotiations on the second sulfur protocol (SSP) have recognized this point, and the agreement will imply different emission reduction rates for the participating countries. A consequence of this is that abatement costs will probably vary more than they would have done if considerations of fairness had played a dominant role in the negotiations. This implies that the economic burdens of abatement, measured per capita or as a percentage of GNP, will vary significantly. Against this background, the question arises as to whether economic instruments have a role to play in the implementation of the new protocol.

The protocol will be signed by the governments of the participating countries. We will therefore regard countries as the basic units, although individual firms may be considered in the future. We will not consider emission charges, which in order to be effective would have to be differentiated across nations, since such a system of charges is an untried form of regulation, and is not a realistic instrument as far as the current protocol negotiations are concerned. Emissions trading, however, has been discussed by the parties to the negotiations, and it is likely that such a scheme will be included in the SSP. It is from this perspective that this chapter has been written.

11.2 What Theory Can Tell Us

Airborne sulfur belongs to a class of nonuniformly dispersing pollutants that causes localized damage when it is deposited. The concept of an optimal solution is obtained by integrated assessment modeling, with the calculation of abatement cost functions, atmospheric transport and deposition as building blocks. The mathematical formulation of the optimization problem was given by Tietenberg (1985):

$$\min \sum_{i=1}^{n} c_i(e_i^0 - e_i), \tag{11.1}$$

given that

$$\sum_{i=1}^{n} a_{ij} e_i \leq D_j^* + b_j, \quad j = 1, \ldots, m,$$

where c_i is cost of emissions abatement from source i; e_i^0 are initial emissions from source i; e_i are emissions after purification from source i; D_j^* is deposition target at receptor j; a_{ij} is the transportation coefficient; amount from source i deposited at receptor j per unit emitted from source i; and b_j is background deposition at receptor j.

A necessary condition for a solution with positive values of e_i is:

$$c'_i - \sum_{j=1}^{m} a_{ij}\lambda_j = 0, \quad i = 1, \ldots, n, \tag{11.2}$$

where c'_i is marginal purification cost of source i; and λ_j is the shadow price on the deposition target D^*_j (positive for binding receptors, zero otherwise).

Due to the nonuniform dispersion of sulfur and the variation in the ability of the receptors to neutralize depositions, marginal purification costs in the optimal solution are generally different; marginal costs at any one source should equal the total marginal "damage" of emitting an additional unit. The expression for "damage" is the shadow prices of the target load constraints weighted by the transportation coefficients. "Damage" is in quotation marks because shadow prices reflect abatement costs and not proper environmental evaluation. The shadow price on a binding constraint show the increase in the total costs of reducing the deposition target by one unit. Receptors with nonbinding deposition constraints, i.e. where depositions are less than target loads, have zero shadow prices. The evaluation of target loads by shadow prices is therefore not an evaluation of the cost of damage in the true sense. Treating environmental objectives as constraints introduces a kind of zero–one evaluation of these constraints. This kind of model-related evaluation must not be confused with valuing environmental variables as such.

The RAINS model (Alcamo *et al.*, 1990) developed at the International Institute for Applied Systems Analysis (IIASA) provides an efficient tool for solving problem (11.1). Policy efficiency and cost efficiency are taken care of by the RAINS model in the following way: Policy objectives are entered as constraints, and cost efficiency is achieved by optimization. The policy objectives are ceilings on depositions, called target loads, in different geographical areas. These geographical areas, called receptors, are 150 × 150 km rectangles into which Europe has been divided. The optimization implies that those levels of emissions are found which minimize total abatement costs summed over all countries, while the depositions at each receptor are within the constraints. Thus, the model assumes policy efficiency with regard to the constraints and solves for cost efficiency.

The type of model represented by RAINS has been used as a frame of reference for the SSP negotiations. There is a general agreement that uniform cutbacks, as under the current protocol, are not cost-efficient, and that concern for the environment should be based, directly or indirectly in some form, on critical loads of sulfur deposition in the grid-based receptors. These critical loads are defined as the amount of deposition in a specified geographical area that is consistent with no significant damage to a given percentage of ecosystems in that area. For instance, the "5% critical load" is defined as the amount of deposition that is consistent with no significant damage to 95% of the ecosystems in a given area. Deposition may be measured in a variety of units, but it is common to measure this in grams of sulfur per km^2 per year. Interested readers are referred to Downing *et al.* (1993) for a comprehensive definition and discussion of this concept.

The recent popularity of emissions trading within countries stems from the certainty of effect, since the total emissions are fixed, and increased cost efficiency results with a high degree of certainty from the economic motivation to trade. The advantages of emissions trading will generally contribute to cost effectiveness and burden sharing, and will also provide a dynamic incentive for more efficient technology choices (see, for example, Førsund, 1992).

Emissions trading has been dealt with in the literature based on models similar to the one above (see Montgomery, 1972, 1974; Krupnick *et al.*, 1983; McGartland and Oates, 1985; Tietenberg, 1985). Looking for decentralized permit systems to mirror market solutions, Montgomery (1972) demonstrated that by pricing permits at each receptor, and forcing a source to acquire permits for all downwind depositions, a decentralized market solution supporting the optimum may exist. Such an approach, however, is probably impossible to enforce in practice. Investigating a system of emission permits for sources, Montgomery (1972, 1974) found arrangements that lead to optimal solutions only in the case where all constraints in (11.1) are binding. Krupnick *et al.* (1983) relaxed this assumption by introducing exchange rates. These rates are defined as the amount of emission reduction at one source required in order to compensate for one unit of increase at another source. The exchange rates discussed by Krupnick *et al.* are endogenous. We feel that a trading system, in order to be feasible, has to be as simple as possible, and in our view this necessitates *exogenous* exchange rates. At this point it is necessary to point out that we are not looking for a second best system, but rather a third or even fourth best, and we will pursue such a system without strict formal criteria.

Emissions trading with exchange rates is not in use even within countries at present. We therefore underline the difficulties within a very simplified system for trade between countries, and conditions on the trading system that we feel must be fulfilled for practical applications of sulfur emissions trading.

11.3 Conditions for Emissions Trading

The SSP will in all likelihood be based on emission ceilings for each country to be achieved by some future date. Since the main rationale for emissions trading is improvement of cost efficiency, there is no reason for introducing trade if emission ceilings are based on the cost-optimal solution with environmental objectives as constraints, nor if the data on which the optimal solution is based are accurate. If some countries are dissatisfied with their emission ceilings, e.g., due to high abatement costs, this problem could be resolved with some system of side payments rather by than moving away from the optimum by allowing emissions trading. It is important to note that given the environmental objectives, considerations of fairness should not interfere with how emission ceilings are set. Such considerations are however valid in the bargaining over how to finance the abatement process (see Vislie and Nævdal, 1993) for a game theoretic discussion of agreements such as the SSP).

In the context of the SSP a number of additional conditions must be fulfilled in order for a trading scheme based on exchange rates to be an attractive policy instrument. At least the following features must be well defined:

- Emission ceilings.
- Environmental policy objectives.
- Exchange rates.

The SSP will probably call for substantial reductions in emissions in almost all countries over the next two decades. Trade implies that countries buying emission quotas will reduce their emissions by amounts lower than those specified in the SSP. One must therefore take special care to argue that an increase in emissions, compared with the ceilings set in the protocol, will constitute an improvement of the SSP agreement. However, even with such a trade, most countries will have to reduce emissions substantially from pre-protocol levels. When trade is allowed, some countries will reduce less than the agreed emission ceilings if they can get other countries to carry out larger reductions.

It is assumed that the new protocol will lay down a quantitative emission ceiling for each country. If this is the only expression of environmental objectives in the agreement, it follows that a one-to-one trading (offsetting an increase in emissions by one country with an equal decrease in another country) is the only possibility. This runs contrary to the recognized nonuniform dispersion of sulfur. Notice that it is not enough to state some general desirability of obtaining critical loads at some future date. Environmental objectives must be formally linked to the emission ceilings in order to define trade ratios other than one-to-one. The formal linkage of the emission ceilings to environmental objectives is through rules for obtaining cost efficiency and policy efficiency. A natural alternative to emission reductions as environmental objectives *per se* is deposition load reductions. The case for this approach, in the context of the SSP, is substantially strengthened by the critical loads concept defined in Section 11.2. This is because the critical loads concept provides a quantitative link between the amount of deposition in different geographical areas and the amount of ecological damage done by the deposition. Although this link measures damage in a physical rather than an economic sense, it could still provide a powerful tool in the formulation of efficient environmental policies.

Assuming that there is agreement on environmental objectives reflecting localized costs, a one-to-one trading scheme cannot be "correct". This is because one-to-one trading would tend to equalize marginal abatement costs, which in general will be inefficient due to the nonuniform dispersion of sulfur emissions.

Since we have assumed that there is a formal link between emission ceilings and environmental objectives, it follows that a change in the distribution of the emissions due to trade will influence the degree to which environmental objectives are fulfilled. Agreement is therefore needed on how to evaluate the impact of trade on the objective variables. We will assume that target loads are the environmental objectives. If the SSP is not perfect, the target loads can be in one of three possible states: (i) the emission ceilings generate depositions that are greater than the target loads; (ii) the emission ceilings imply that depositions are less than target loads; and (iii) some depositions are below target loads, and some are above.

In the first two cases, emission ceilings are generally too high or too low. It follows that in these cases emissions trading with exchange rates is not a efficient instrument since emission ceilings are not consistent with environmental objectives. The relevant strategy in this case is either to renegotiate the emission ceilings, or to rethink the environmental objectives. If this is not done, we as may well implement a one-to-one trading scheme in order to achieve the total level of emissions at the lowest possible cost.

If depositions are both below and above target loads, there is room for improvement in the distribution of emissions, and emissions trading with exchange rates may be a suitable instrument. It is important to note that there is no guarantee that all deposition targets will be met after trading has taken place. Ideally this would be the case, but if target loads are violated before trading has taken place, we will not be able to achieve nonviolation for all target loads with certainty unless the initial emission ceilings are renegotiated. [Van Ierland *et al.*, (1993) achieve such a result through one-to-one trading, using a procedure that assumes that the sum of the initial emissions is equal to the optimal emission level. They also allow forced trading in the sense that trading takes place which would not occur in the real world without additional compensation to the traders.]

A necessary condition for establishing appropriate trade ratios is to calculate a reference solution, covering the distribution of emissions and abatement costs, that must be agreed upon. The logical choice for such a reference solution would be the optimal solution derived from model calculations based on environmental objectives, and whatever imperfect information we have at our disposal. In the absence of such a solution, one-to-one trading seems the only logical trading scheme, since no reference solution implies no consistent policy objectives that acknowledge the physical properties of sulfur emissions.

We are now left with the difficult task of defining meaningful exchange rates. One possible way of defining a set of exchange rates that is not based on the cost minimization model, is to base them on ratios of "ecological damage", derived from the matrix describing the dispersion of sulfur and some kind of evaluation of the tolerance of the receptors to sulfur depositions. The difficulty with such an approach lies in the evaluation of damage to the receptors. To establish universally acceptable criteria for such an evaluation seems a difficult task as long as one is outside the realm of pure economic theory, where it is at least assumed that it is possible to evaluate the benefits. In our case, it would require the establishment of benefit functions, one for each receptor, that would have to be established according to criteria that will be acceptable to all signatories to the SSP. We feel that the alternative presented below is better in the sense that it is simpler and still maintains some beneficial properties.

In this chapter we base the exchange rates on the information contained in the first-order conditions (see Klaassen *et al.*, 1994). If we divide the first-order conditions with one another we obtain the expression:

$$\gamma_{ij} = \frac{c_i'}{c_j'} = \frac{\sum_{k=1}^{m} a_{ik}\lambda_k}{\sum_{k=1}^{m} a_{jk}\lambda_k}. \tag{11.3}$$

We define γ_{ij}, evaluated in the optimum, as the exchange rate between country i and country j; that is, one unit of emissions from source i is equivalent to γ_{ij} units of emissions from source j. The choice of this exchange rate may seem somewhat arbitrary, in the sense that its information is of limited relevance outside the optimal solution. However the scanty information that exists will give the countries an incentive to trade emissions until they are on a point on a isocost curve where the curvature is the same as in the optimal solution. This trading will take some account of the environmental impact of the changed distribution of emissions. If country i contributes relatively more to binding receptors, then country i may increase its emissions by less than one unit for every unit of emissions it buys from country j. If the opposite is the case and country j contributes relatively more to binding receptors, then country i may increase its emissions by more than one unit for every unit of emissions it buys from country j. The result may be considerable cost savings without altering the distribution of emissions in a environmentally unfavorable manner. It should be clear, however, that if emissions trading with exchange rates results in exactly the optimal solution, then this is to a large extent due to plain luck. The point is that trading with exchange rates has the potential to reduce the gap between optimal emissions and actual emissions to a considerable degree. The conditions under which this will be the case are examined in Section 11.5.

11.4 Administrative Framework

There must be some administrative framework within which trades between countries can be organized, in order to capture the advantages of emissions trading.

The motivation for countries to trade emission permits is taken to be the desire to save abatement costs. There is of course the possibility that some countries with a high level of environmental consciousness would use trading as a means of financing emissions abatement to a level below the emission ceilings specified in the reference solution, but we will disregard this possibility in the following. A number of issues have to be dealt with:

- How to introduce exchange rates for sulfur emissions.
- How to ensure that trades are in compliance with the protocol.

- How to approve trades.
- The roles of third parties.

As a point of departure, we assume that the administration of the protocol, the UN ECE (United Nations Economic Commission for Europe), provides a *service function* with respect to the reference run and calculations necessary to monitor the consequences of trades on the level of deposition. We refer to this function as the *Service Center.*

The emission targets are to be met by a specified future date, say 10 years ahead. This means that the trades do not need to be carried out immediately; rather, the point is to agree on future commitments. The trade proposals must be accepted by some evaluation procedure to be a valid part of the agreement, and given the time span of the SSP, there should be ample time to evaluate proposed bilateral trades.

There are two opposite extremes for institutional setups: centralized command, and decentralized consensus. These administrative schemes are illustrated in *Figure 11.1.*

11.4.1 Centralized framework

The Service Center will be upgraded to a *Command Center* in the centralized solution. The trades are organized and evaluated by the Center, but it is left to the countries to actually carry them out. The objective of the Center is to find trades such that total costs are reduced.

A key feature in the centralized solution to trade is how to deal with the environmental objectives. If the target loads are to be observed, all trades will be interlinked, and all trades must be determined in a simultaneous decision process. Thus, if applying a sequential trade procedure, the order of trades will influence the total outcome. When a target load is reached new trades will be decisively restricted.

Working with two periods may help to resolve this problem by registering all feasible combinations of trade out of the set of trades cleared by the countries involved, and then solving simultaneously toward the end of the trial period.

One should note that the need to introduce the target load constraints explicitly is not necessarily so strong. Recall that a subset of target loads is more than fulfilled in the initial protocol agreement. There is therefore some scope to increase emissions without violating the target load constraints. An *ad hoc* procedure would be to check for violations and rule out trades that lead to excessive depositions also in third countries, or to restrict the volume of such trades.

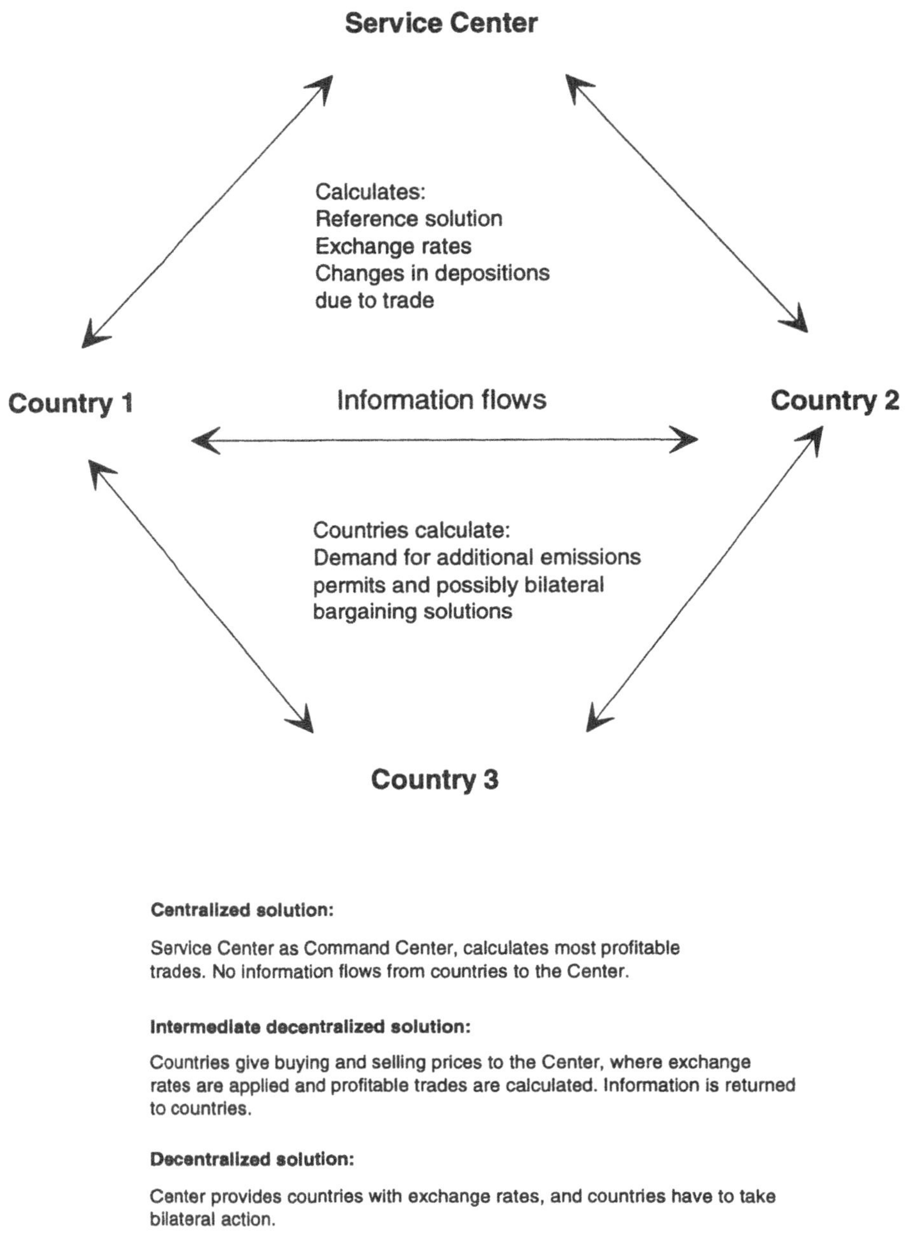

Figure 11.1. Illustration of information flows for different institutional frameworks.

11.4.2 Decentralized framework

In a decentralized framework the countries must take some initiative by proposing trades in the trial period. One question is how countries could identify possible trades. It could be left up to every country to seek out all other countries and exchange bids and offers, as indicated in the third setup in *Figure 11.1.* This approach does not appear too promising. It would be better to use the Service Center as a clearing house by receiving bids for selling purification capacity and receiving offer prices for buying emission quotas (intermediate decentralized solution). If some countries feel that purification costs are excessive, they could quote what they are willing to pay to other countries to assume responsibility for a certain level of abatement. Similarly, if countries feel that they can purify even more without burdening themselves with excessive costs, they could quote how much they would accept as payment to undertake more purification. Countries could quote on both types of bids simultaneously. The countries do not have to consider the exchange rates or trade ratios when making bids and offers. By applying exchange rates, the Service Center can delineate the set of possible trades and provide countries with appropriate information upon which they can act. For example, weights of sulfur emissions could be converted to a common "damage" unit, i.e., by applying exchange rates; thus emissions from Norway and Turkey can be compared on a bilateral basis.

Since no changes in deposition have yet to take place, there is time both for countries to work out the most profitable trades, and for the Service Center to check the target load constraints. When these calculations are completed and possibly subjected to some kind of approval procedure, trading may take place and emission permits may change hands.

11.4.3 Zoning

It is undoubtedly easier to decentralize emissions trading when exchange rates are all equal to one. In order to benefit to some extent from this, and to reduce the number of trade ratios, one could consider some sort of zoning. One possibility with the greatest flexibility is to restrict trade to neighbors. Countries could also be grouped according to the level of target loads and downstream – upstream features of atmospheric transportation. Trades within a zone would then be on a one-to-one basis, while exchange rates apply to intergroup trades. The Service Center would function as before.

11.5 Application of the Trade Algorithm SLEETS

To study the properties of exchange rate trading, we have developed the computer model SLEETS. A main priority has been to simulate as closely as possible how trading might take place in the real world. The program takes as inputs two emission vectors: one initial solution which is the starting point of the algorithm, and one optimal solution which is the one we want to attain. The initial solution may be interpreted as an emission vector that has been accomplished through the UN ECE bargaining process. The optimal solution is generated by the RAINS model version 6.0, using the target loads currently recognized in the UN ECE negotiations. Associated with these vectors is a distribution of abatement costs, generated by exponential abatement cost functions. We have based these functions on the cost functions generated by the ENEM module in the RAINS program. The model then proceeds to examine the possible cost savings of the pairs of bilateral intercountry cooperation. The pair that may realize the largest cost savings is chosen and emissions between these countries are reallocated in a cost-minimizing manner. Thereafter, all possible cost savings through bilateral cooperation are recalculated, and again the most profitable pair is chosen. This continues until cost savings are smaller than a predetermined level, after which the algorithm ends.

We concentrate on three scenarios: the near-optimal scenario; the significant variations scenario; and the significant exceedances scenario.

11.5.1 The near-optimal scenario

This scenario is intended to simulate the event that the emission vector resulting from the bargaining process is very close to the optimal solution. This scenario has been generated by letting the initial emission vector vary ±20% around the optimal solution. The main purpose of this scenario is to examine whether an exchange rate trading system will ruin a "rather good" bargaining result. The aggregated results from this scenario are given in *Figure 11.2.* It is evident from this figure that exchange rate trading has next to no impact on the aggregate figures. Cost savings are in the neighborhood of DM 500 million per year, and total emissions are reduced by 109 kt SO_2. This is rather promising. The proximity of the initial solution to the optimal would make large changes in the aggregate figures due to undesirable trading. However, these figures are not the main point. The basic test of exchange rate emissions trading is the effect on the *distribution*

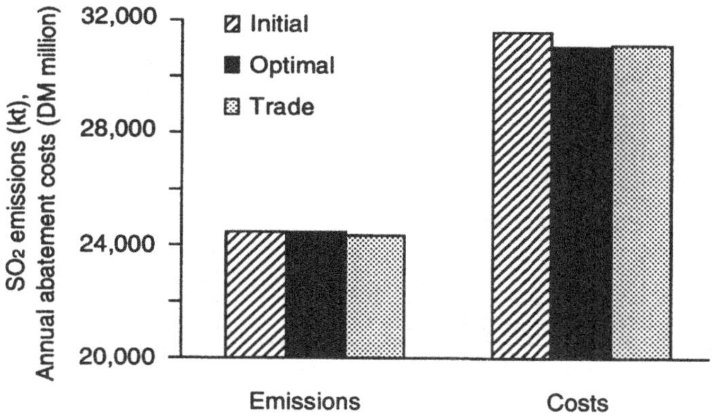

Figure 11.2. The near-optimal scenario.

of costs and emissions. For this purpose we have created a simple index. For all countries we compute:

$$\sigma = \frac{x_{\text{trd}} - x_{\text{opt}}}{x_{\text{ini}} - x_{\text{opt}}} \tag{11.4}$$

where x_{ini} is the initial emissions before any trading has taken place, x_{opt} is the country's optimal emissions and x_{trd} is the country's emissions after all trading has taken place. The properties of this index should be evident. If x_{trd} is closer to x_{opt} than x_{ini}, then the value is less than 1. A negative sign implies that trading has resulted in overshoot. The best possible result is $\sigma = 0$. If $\sigma = 1$, the country's emissions are either unaltered by the sequence of trades, or, most likely, the country has not engaged in trading at all. The values of these indices are illustrated for the near-optimal scenario in *Figure 11.3.* Note that we have left out the countries that did not trade.

As should be evident from *Figure 11.3*, all countries that engage in trading are closer to the optimal solution after trading than before. This result might seem somewhat optimistic, but this property is pervasive in our simulations, although a few exceptions have occurred. All in all, however, it seems that exchange rate trading has the fortunate property that it does not significantly worsen an initial solution that is close to the optimum.

11.5.2 The significant variations scenario

This scenario is generated by increasing the optimal emission vector so that the resulting initial emissions are uniformly distributed ±60% around the optimal solution. We consider this scenario an important test of exchange

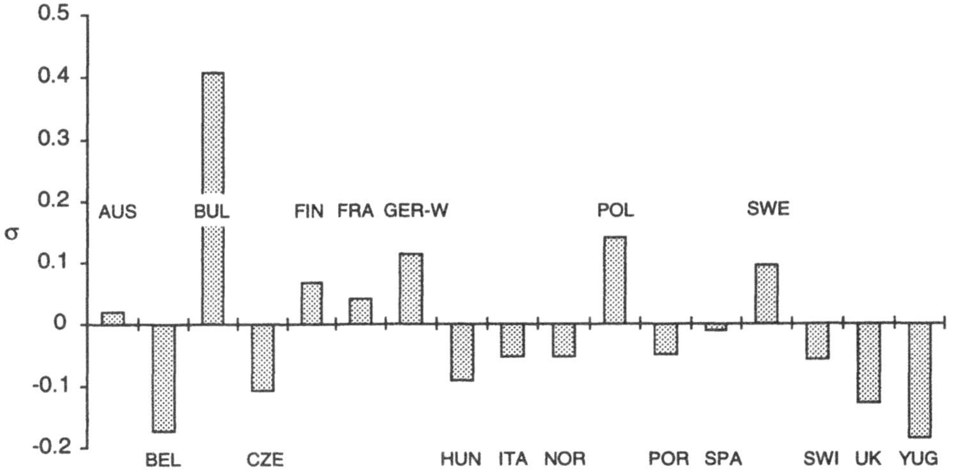

Figure 11.3. The near-optimal scenario: trading indices.

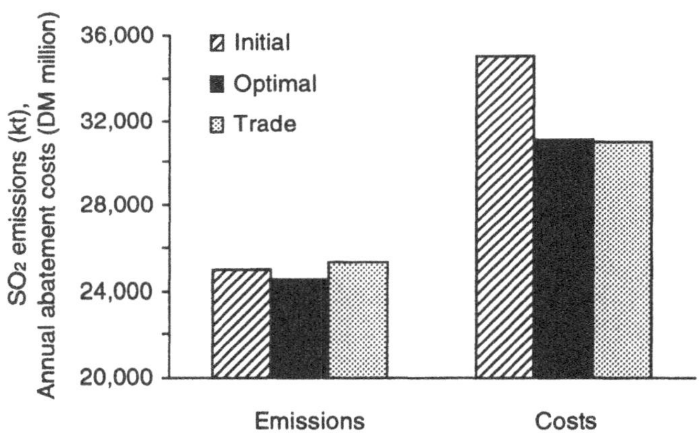

Figure 11.4. The significant variations scenario.

rate trading. The potential for cost savings is large, and if a considerable amount of these savings are realized through exchange rate trading, then this system deserves to be taken into consideration. The aggregate figures of this scenario are presented in *Figure 11.4.* The cost savings are considerable – almost DM 4 billion per year and have occurred without any significant increases in aggregate emissions (+290 kt SO_2).

But the important thing is still the distribution of emissions. Using the index defined in the previous section, these are shown in *Figure 11.5.* Again

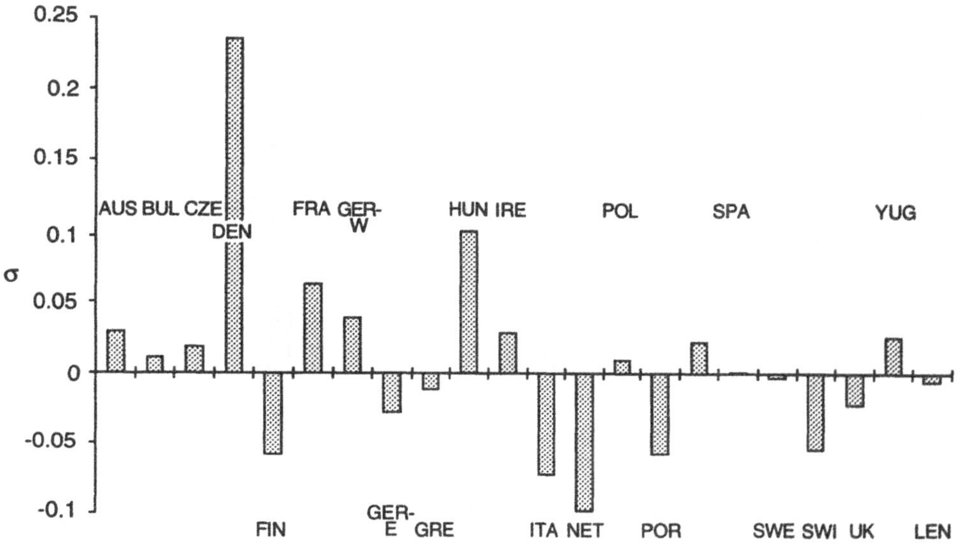

Figure 11.5. The significant variations scenario: trading indices.

we have left out the countries that did not participate in any trading. As one can see, the simulation is almost suspiciously successful. All countries engaged in trading are considerably closer to optimum. Spain, for instance, emitted 2,209 kt SO_2 before trading, compared with 1,487 kt after trading, measured against an optimal level of 1,486 kt. Denmark is the country where σ is the largest. The optimal level of Danish emissions is 58 kt. Trading increases the emission level from 27 to 51 kt, which from an efficiency point of view must be considered an improvement.

11.5.3 The significant exceedances scenario

In this scenario all countries have been given emission increases that are uniformly distributed from 0 to +30% above the optimal solution. This is a somewhat different scenario from the preceding ones in that all emissions are higher than optimal. Under these circumstances one cannot expect a solution close to the optimal solution. This is confirmed by *Figure 11.6*. There is a small decrease in aggregate emissions (approximately 300 kt) and the cost savings are DM 247 million. As in the preceding scenarios, we have constructed trade indices for the countries that participate in trading. From *Figure 11.7* we see that the effect on the distribution of emissions is

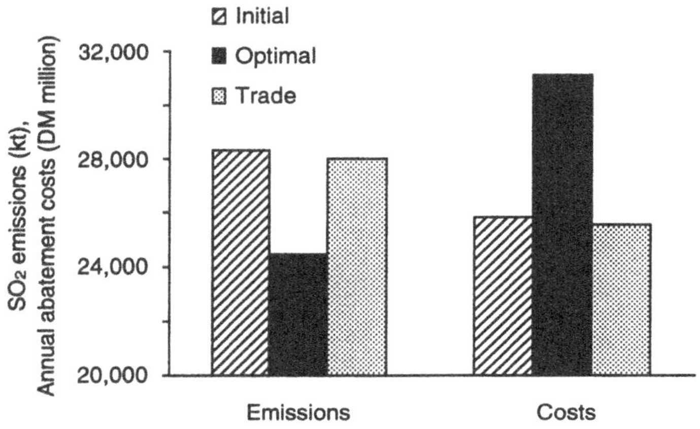

Figure 11.6. The significant exceedances scenario.

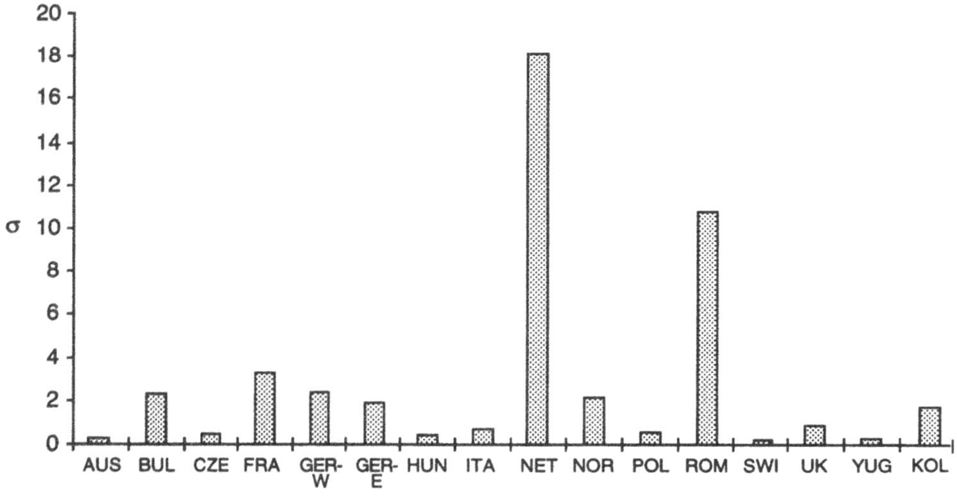

Figure 11.7. The significant exceedances scenario: trading indices.

not very good. Although some countries come closer to the optimum, others move further away, particularly the Netherlands and Romania.

In this case, emissions trading with exchange rates has somewhat disappointing results. This is not because something is wrong with the concept, but rather because trading started from an initial emission vector that was obviously built on a different evaluation of environmental objectives. That is, the reference scenario we tried to reach was not consistent with the emission ceilings required in the initial solution. This implies that we started trading with the wrong set of exchange rates.

11.6 Conclusions

Emissions trading has the potential to combine both policy efficiency and cost efficiency. To date, trading schemes with one-to-one trading ratios have been set up. However, sulfur emissions disperse nonuniformly and therefore bilateral, different exchange rates for emissions trading are required. A number of conditions must be agreed upon within the SSP if emissions trading is to function. Most important, agreements are required on environmental objectives, a reference solution, generation of exchange rates, or definitions of how to evaluate changes in depositions following trade.

A general feature of airborne sulfur as a pollutant is that a unit emitted from a source has different impacts depending on where it falls. This implies that one unit of increase from one source cannot be offset by exactly one unit decrease from another source, but has to be weighted by some kind of exchange rate. The exchange rate may be greater or smaller than one. We have suggested one way of defining such exchange rates, and pointed out that these have some beneficial properties. Emissions trading with different bilateral exchange rates is a demanding task for implementation, and such systems are not yet in use, even within countries. We have underlined some of the difficulties and conditions that must be met for practical applications of sulfur emissions trading.

The simulations of three scenarios of sequential bilateral trade have been done with SLEETS. The work so far has led to the following conclusions. Trading with exchange rates seems beneficial if aggregate emissions at the starting point are close to the aggregate emissions at the optimum. The cost savings that may be achieved are significant. If aggregate emissions initially are considerably higher or lower than in the optimum, the method seems to be less appropriate.

Acknowledgments

The authors wish to thank Ger Klaassen and two anonymous referees for comments and suggestions that significantly improved this chapter. We also wish to thank the Norwegian Ministry of Environment and the Norwegian Ministry of Finance for making funds available so that this article could be written.

References

Alcamo, J., Hordijk, L., and Shaw, R., eds., 1990, *The RAINS Model of Acidification*, Kluwer Academic Publishers, Dordrecht/Boston/London.

Førsund, F.R., 1992, Emissions Trading, Paper presented at the First Meeting of the Task Force on Economic Aspects of Abatement Strategies, 2–3 March, UN ECE, Geneva, Switzerland.

Downing, R.J., Hettelingh, J.-P., and de Smet, P.A.M., eds,. 1993, *Calculation and Mapping of Critical Loads in Europe: Status Report 1993*, National Institute of Public Health and Environmental Protection, Bilthoven, Netherlands.

Klaassen, G.A.J. , Førsund, F.R., and Amann, M., 1994, Emissions Trading in Europe with an Exchange Rate, *Environmental and Resource Economics*, forthcoming.

Krupnick, A.J. , Oates, W.E., and van de Verg, E., 1983, On Marketable Air Pollution Permits: The Case for a System of Pollution Offsets, *Journal of Environmental Economics and Management* **10**:233–247.

McGartland, A.M., and Oates, W.E., 1985, Marketable Permits for the Prevention of Environmental Deterioration, *Journal of Environmental Economics and Management* **12**:207–228.

Montgomery, W.D., 1972, Markets in Licenses and Efficient Pollution Control Programs, *Journal of Economic Theory* **5**:395–418.

Montgomery, W.D., 1974, Artificial Markets and the Theory of Games, *Public Choice* **18**:25–40.

Tietenberg, T.H., 1985, *Emissions Trading: An Exercise in Reforming Pollution Policy*, Resources for the Future, Washington, DC, USA.

Van Ierland, E.C., Kruitwagen, S., and Hendrix, E.M.T, 1993, Tradable Discharge Permits for Acidifying Emissions in Europe: 'Guided' Bilateral Trade, Paper presented at the International Conference on Economic instruments for Air Pollution Control, 18–20 October, International Institute for Applied Systems Analysis, Laxenburg, Austria.

Vislie, J., and Nævdal, E., 1993, International Agreements on Nonuniformly Dispersed Pollutants, Paper presented at the International Conference on Economic Instruments for Air Pollution Control, 18–20 October, International Institute for Applied Systems Analysis, Laxenburg, Austria.

Chapter 12

Agency in International Pollution Permit Trading

Dallas Burtraw
Quality of the Environment Division
Resources for the Future, Washington, DC, USA

Abstract

This chapter investigates institutional issues affecting the potential design of an international pollution permit trading program. Most previous studies, including simulation models of SO_2 emissions trading in Europe, have assumed that national governments would be the agents participating in trades. An alternative is that emission permits be distributed to and traded by affected enterprises. This chapter evaluates these two alternatives according to the two primary benefits usually attributed to incentive-based environmental policies. The first is the attainment of cost effectiveness, also referred to as technical or productive efficiency. The second is the internalization of social costs, otherwise known as allocative efficiency. I conclude that an international system of emission permit trading such as SO_2 trading in Europe, would perform better according to these economic criteria if permits were distributed directly to affected enterprises. However, in this brief investigation, such a conclusion can not be definitive. Rather, the primary finding is that further attention to the question of agency is warranted. Nevertheless, a strong case can be made that international environmental agreements such as an accord for SO_2 emissions trading in Europe should establish guidelines for national governments regarding the domestic economic regulation of internationally tradable permits.

Key words: sulfur dioxide, emissions trading, principal–agent, international environmental negotiation.

12.1 Introduction

Incentive-based approaches to environmental regulation are moving into the spotlight in many industrialized nations. The movement is occurring through a broad, albeit so far limited, array of applications addressing local pollution issues and through increasingly sophisticated discussions concerning transboundary pollutants (OECD, 1993). Incentive-based (IB) approaches are distinguished from the conventional approach to environmental regulation, generally labeled command-and-control (CAC), which typically involves the specification of design or performance standards for given technologies in given applications. The hallmark of IB approaches is that individual firms are given latitude in design and performance, while they are held accountable for their consumption of services provided by the environment. Accountability is enforced either through quantity constraints on total collective emissions or on ambient air quality, or through price constraints such as fees per unit of emission of various pollutants.

This chapter addresses the question of what level of organization – economic enterprises or national governments – should be the institution designated as the decision making unit for environmental compliance strategies, when IB instruments are adopted on an international level for the regulation of transboundary pollutants. For instance, under an international system of tradable emission permits negotiated on a multilateral basis, the question at hand is to which type of institution should accrue the internationally created intangible property rights associated with emission permits. Should permits be distributed to and traded by economic enterprises or national governments? I characterize this question as the designation of agency, with the particular context for this discussion being negotiations for sulfur dioxide (SO_2) reduction in Europe.

Most discussions of IB regulation at the international level have been either silent or ambivalent on this question. The implicit assumption is most often that national governments constitute the economic agents that would engage in permit trading (OECD, 1992; UN, 1992). One justification for this is that political acceptance of a tradable permit system may be most forthcoming if it is perceived to be a relatively small departure from conventional multiparty negotiations that assign targets among nations, such as recent agreements on SO_2 reductions in Europe. Trading between nations would be a mechanism allowing flexibility in the implementation of these targets.

Furthermore, national governments are indisputably the sole sovereign authority empowered to enter into international environmental agreements;

and therefore, national governments are the *principals* ultimately accountable for compliance with and enforcement of those agreements. This status makes national governments the natural candidates to function also as the participating *agents* in a permit trading system.

A further reason for national governments to be the agents in an international system of permit trading is that they are in a position to integrate standards for compliance with domestic environmental regulation. This is often referred to as harmonization of international environmental obligations with domestic regulation, and is especially relevant if nations already use IB environmental regulation to protect the environment or to raise revenues. Hoel (1993, 1991a) argues that to designate national governments which are the principals to an agreement to serve also as the agents for its implementation is the best design for the harmonization of policies and the maximization of social welfare from an international perspective.

An explicit assumption in recent simulation modeling of potential agreements for SO_2 emission reductions in Europe has been that national governments serve as the agents for trading. For instance, the RAINS model developed at the International Institute for Applied Systems Analysis (IIASA) is only designed to entertain policy options including emission constraints targeted at national governments. RAINS uses cost functions for emission reductions aggregated by industry and then aggregated by country to reach a solution about the economic and environmental cost of potential policies. Based on the summary reported in UN ECE (1992), this statement also holds with respect to the Coordinated Abatement Strategy Model (CASM) developed by the Stockholm Environment Institute at York, and the Abatement Strategies Assessment Model (ASAM) currently being developed at Imperial College, London.

However, several authors have suggested that trading between nations may not be the only, or the best, design for an international permit trading system. Swart (1992) points to lack of experience by national governments, lack of knowledge about compliance options and operations at the facility level, and the inability of bureaucracies to act in a timely manner in a market environment. He suggests that greater cost effectiveness might be achieved by trading between firms, although he acknowledges this would imply transfers of currency that may conflict with national trade policies and therefore permit trades might have to be ratified by the national governments involved. Grubb (1992) suggests that governments might pass entitlements to permits obtained through international agreements on to firms, which could engage in trading. Bertram (1992) considers an assortment of decentralized approaches such as allocating permits on a per capita basis. Dudek and Tietenberg (1992) argue on practical grounds that the issue would have to be addressed by each nation. In capitalist economies, they suggest it is most plausible for firms to serve as agents for permit trading, while in centrally planned economies central authorities would engage in trading. They point out that the Montreal Protocol, as amended in London, allows trading between enterprises subject to the concurrence of national governments that are involved. And they point out that trading of SO_2 allowances under the 1990 Clean Air Act amendments in the USA is between affected firms, although they do not draw attention to the pivotal role that economic regulators at the state level have in designing cost recovery rules, approving trades, and ultimately affecting the shape of the program (Bohi and Burtraw, 1992, 1991).

The choice between enterprises versus national governments as the authorized trading agents in an international permit system represents conceptually different regulatory frameworks. In any case, the choice is not a clear-cut one, because if enterprises constitute the agents for trading, national governments or the relevant economic regulators will retain overarching authority with regard to policies governing the economic behavior of enterprises (much as state regulators retain such authority with regard to permit trading by utilities in the USA). However, in principle, international agreements could incorporate prescriptions or guidelines about the proper role for national governments or economic regulators in regulating firm behavior with regard to compliance under an international permit system. These types of guidelines would seem more likely to be found in an agreement in Europe for control of sulfur deposition than in broader international environmental agreements because of the pre-existing legal framework within the EU and the apparent desire of many central and eastern European countries to move toward greater economic integration with the EU.

This chapter employs two economic criteria to address the question of agency in a international permit system, particularly with regard to SO_2 emission reductions in Europe. These criteria are the two significant potential benefits that are generally ascribed to IB environmental regulation. The first is the attainment of cost effectiveness, also referred to as technical or productive efficiency. Section 12.2 evaluates the two alternative approaches with regard to their ability to promote the cost-effective implementation of a given environmental goal and to provide dynamic incentives for technological advance. The second benefit ascribed to IB environmental regulation is the internalization of social costs, otherwise known as allocative efficiency. Although these two anticipated benefits of IB policies are often thought of as two sides of the same coin, in practice they have independent relevance. Furthermore, most conventional approaches to regulation that have emphasized cost effectiveness have ignored the internalization of social costs. Section 12.3 evaluates the alternative possibilities with regard to the internalization of social costs and transparent pricing of factors of production, that is, prices that send accurate signals about the social cost of resource use. Finally, Section 12.4 presents a summary and offers a perspective on the role of government in administration and monitoring.

12.2 Comparing Performance in Promoting Productive Efficiency

12.2.1 Introduction

The most commonly cited potential benefit of IB regulation is the attainment of productive efficiency through the cost-effective implementation of a given environmental goal, along with the dynamic incentive for technological advance in order to reduce the cost of compliance in the future. Formally, within a theoretical model of a competitive market in which all parties have full information, tradable permits or emission fees can achieve a given environmental goal at lower cost than any other regulatory scheme (Montgomery, 1972; Baumol and Oates, 1988). Alternatively stated, IB approaches can afford society a greater level of environmental quality at the same cost as might be expended with CAC regulation. Furthermore, the flexibility in the choice of abatement strategy imparts an incentive to improve abatement technology because some portion of the savings will accrue to the firm.

Several studies suggest that a system of tradable SO_2 emission permits in Europe has the potential to achieve dramatic cost savings relative to rigid and inflexible assignments of emission quotas on a national basis. Simulations based on the RAINS model indicate that a reallocation of the current level of expenditures between nations in Europe, as might be achieved through a cost-effective system of permit trading among national governments, could further improve the emission reductions that will be achieved from 18%–30% of 1990 levels. However, reductions in sulfur emissions and deposition levels by 50%–70% is required to reverse the trend of soil acidification in Europe. However, if this effort was not uniform but was targeted, this goal could be obtained with reductions in emissions of 20%–25% (Shaw *et al.*, 1990, p. 352).

An international agreement for SO_2 emission permit trading that casts national governments as the agents for trading may capture a significant portion of potential cost savings compared to traditional approaches to environmental regulation. However, this chapter suggests that there is significant reason to suspect that such an approach would fall short of the total potential for savings. In this light, the relevant question is whether a system that casts national governments in the trading role would perform any differently than if affected economic enterprises were cast in that role.

There are at least two reasons to think they would perform differently. One reason is that national governments may fail to adopt domestic policies that would implement their international obligations in a cost-effective manner. A second reason has to do with the mechanics of emissions trading when spatial considerations are important. These issues are discussed in turn.

12.2.2 The shape of domestic policies under an international agreement

If national governments were designated the trading agents in an international SO_2 permit trading program, each government would have to develop a domestic compliance plan. The political process involved in the development of a compliance plan would remain under the influence of competing social priorities and private interests. As a consequence, it is not clear what form domestic environmental regulation would take.

If an international environmental agreement took the form of an international tax imposed on national governments, it seems reasonable to expect that each national government would pass the revenue burden along to polluters in the form of a domestic emission tax in order to raise revenue to fulfill its international fiscal obligation (Hoel, 1993). However, for a variety of reasons, it appears that tradable permits are the likely instrument of choice at the international level, at least with respect to the control of SO_2 emissions in Europe. Unlike a tax system, under a permit system national governments would not be saddled with large requirements to generate revenues.

However, national governments would be saddled with an emission cap. To comply with the cap these governments would have latitude to consider passing on permits to affected enterprises directly. But they may as well adopt domestic tax or permit programs, or extend and calibrate existing CAC regulations in order to comply with their international obligations.

The latter is the smallest departure from the status quo, and is perhaps the most likely to occur.

The evolution of emission offset and netting programs in the USA provides an interesting example of the behavior of regulators when faced with an emission constraint on an aggregate level, absent well defined institutions to support an innovative approach such as emissions trading. With the development of the US Clean Air Act, many metropolitan areas were designated as being in noncompliance with National Ambient Air Quality Standards (NAAQS). The responsibility for regulating emissions and ultimately improving air quality is delegated primarily to the states, which are required to develop and implement state implementation plans to bring metropolitan areas into compliance with NAAQS. In many instances metropolitan areas are subject to a virtual emission cap, which primarily affects point sources of pollution, that is enforced by sanctions imposed by the federal government. Meanwhile, the desire for economic growth brought with it the specter of new point and mobile sources of pollution that further undermined local air quality.

One result of this dilemma was the introduction of emission offset and netting programs to accommodate new growth without worsening ambient air quality. Static savings resulting from regulation of conventional air pollutants in the USA have been estimated at hundreds of millions of dollars (Hahn and Hester, 1989; Tietenberg, 1990). However, as has been discussed thoroughly elsewhere, these programs were often flawed and trading was occasional, partly due to the imprecise and uncertain allocation of property rights (Tripp and Dudek, 1989; Tietenberg, 1985; Hahn, 1989; Dudek and Palmisano, 1988). The perceived lack of commitment to incentive-based approaches by federal regulators bred uncertainty regarding the longevity of the program and of property rights or related entitlements to savings generated through specific investments. In any case, institutions to promote emissions trading have been slow to evolve and thus have had limited success, even in the face of quantity constraints on total emissions. The significant inertia associated with the CAC approach has kept it the centerpiece of State Implementation Plan (SIP) efforts in the USA, even in the face of ostensible emission caps at the metropolitan or regional level.

An international permit trading system that identifies national governments as trading agents would be analogous in structure to the CAC because it would impose an emission cap at the national level, even though trading would allow flexibility in setting the cap. As such, this approach may have an analogous result. The practical appearance of domestic regulation is likely to resemble the status quo, a cost-effective outcome would not be achieved at the domestic level, and the opportunity cost of emissions would not be recognized by firms.

Whether an international permit system with enterprises acting as the trading agents would succeed further in this dimension is an open question. National governments would still have opportunities to influence compliance activities by firms and to strategically subsidize domestic industries if permits are allocated directly to and traded by individual enterprises. Among the most significant of these opportunities applies to regulation of the electricity sector in Europe, which accounts for approximately 65% of total SO_2 emissions in Europe.

Within the electricity industry across Europe there exist a variety of ownership and regulatory structures. Investor-owned utilities regulated at the national or regional level, municipal and cooperative utilities, state-owned enterprise, recent attempts at privatization in the UK and a number of cases where these structures exist side by side in mixed ownership systems with yardstick competition (Henney, 1992; Holmes, 1988).

IB approaches to environmental policy depend fundamentally on the expectation that firms will respond to incentives by choosing a least-cost strategy for compliance. Virtually all of the economic analyses of potential savings from IB regulation have been limited to examinations of competitive product markets, or occasionally, to the role of market power on behalf of producers in an unregulated product market (Baumol and Oates, 1988; Tietenberg, 1985; Bohm, 1981; Hahn, 1984; Malueg, 1990; and Misiolek and Elder, 1989). However, regulated and state-owned enterprises that characterize the European electricity sector may not have the incentive to maximize profits or to minimize costs that competitive firms have (Henney, 1992, p. 76; Holmes, 1988). In addition, regulated firms typically do not make decisions on the basis of market prices, but rather on the basis of distorted opportunity costs that reflect regulatory practice. Rules governing the recovery of costs for permit expenses and how these compare to rules for alternative investment options would strongly influence the financial incentives facing firms in compliance with a permit trading program (Bohi and Burtraw, 1991, 1992).

Across Europe one currently observes the strong influence of regulatory policy in creating incentives for electric utilities, a common practice by governments to pursue diverse social goals other than economic efficiency, and tremendous dissimilarities in these social goals between countries and in the economic and regulatory structure of electricity systems. These characteristics of the electricity industry represent formidable obstacles to the attainment of cost effectiveness in a permit trading program, but some remedies exist for these problems.

One remedy is to promote the symmetric treatment of compliance options with respect to cost recovery rules and associated regulatory incentives. Another is to promote the EU goal of transparent pricing of electricity. The USA has the Uniform System of Accounts that reconciles the policies of different regulators in a consistent manner. As the EU moves toward an open internal energy market and greater economic integration with the rest of Europe a similar institution would be very helpful. Pertaining directly to tradable emission permits, there will be a number of accounting and regulatory issues that will be new for the industry. A template for consistent accounting should be developed as part of an international agreement for emissions trading.

In summary, it is not obvious that the achievement of potential cost savings would be any more certain if enterprises were to be directly endowed with permits in an international permit system than if national governments took that role. However, this approach would guarantee that a framework for permit trading would be put in place at the domestic level, and it seems likely that, at least at the margin, firms would recognize some portion of the opportunity cost for marginal permit acquisitions. This would not occur if they were regulated solely by CAC regulations under domestic statutes sufficiently stringent to comply with international agreements. In either case, national governments retain the authority and capability to influence the compliance behavior of firms. The main point that emerges, however, is that placing permits directly into the hands of affected enterprises removes one obvious avenue of political influence, and allows the realization of cost savings to remain a possibility.

12.2.3 The mechanics of emissions trading with spatial constraints

An important feature of acidification in Europe is that damages are thought to relate to SO_2 emissions in a nonlinear and nonuniform manner. The relationship between SO_2 emissions and the quantity of sulfur deposition is approximately linear. However, SO_2 emissions are not uniformly mixing in the atmosphere, meaning that the location of sulfur deposition and associated damages depends on the location of the emissions.

The SO_2 emissions trading program adopted in the USA in 1990 ignores entirely the spatial dimension of the acidification problem and trading of SO_2 emission "allowances" may occur between affected facilities located anywhere in the country. Kete (1992) and Hausker (1992) offer compelling defenses for why this was a reasonable policy decision. However, in Europe the spatial dimension of the acidification problem is less likely to be ignored. Current and emerging patterns of industrial activity, coupled with meteorological patterns and the location of sensitive ecological systems, suggest that the migration of emissions through permit trading would not be expected to lessen the problem in acutely affected areas. Moreover, the political context is defined by strong national identities and by a decade of scientific investigation that has heightened awareness of regional effects. Consequently, the promise of IB regulation must be evaluated in the context of a spatially differentiated program incorporating ambient constraints on deposition, with its associated complexities.

A number of approaches to incorporate spatial constraints into a system of emissions trading are possible, which are reviewed in Tietenberg (1985) and Klaassen and Amann (1992). The approach that has received the most attention is emissions trading subject to constraints on deposition of sulfur at various receptor sites, sometimes termed a "pollution offset program." Mathematical programming can be employed to identify the cost-minimizing outcome under such a trading system. The notion that this outcome could be achieved through permit trading rests on the incentive compatibility of the outcome, because it is, in effect, a competitive equilibrium. However, the actual process through which participants move toward this outcome is usually left unspecified. The implicit assumption is that trading between firms follows a tatonnement process facilitated by a "Walrasian auctioneer" in competitive price theory. Mathematically, this is equivalent to the assumption that no trade takes place until an equilibrium is discovered, at which time permits are reallocated on a simultaneous and multilateral basis. Theory suggests the equilibrium will be the unique cost-minimizing outcome.

Beginning with Atkinson and Tietenberg (1991), several recent studies have introduced doubt about the ability of a system of tradable permits to achieve the cost minimizing outcome in practice when permits are traded under a system that accounts for spatially differentiated effects of pollution. In place of mathematical programming, these studies have used simulations of actual trading processes to illustrate trading not as simultaneous and multilateral, but sequential and bilateral instead. In the presence of various trading rules such as an offset program that requires that individual trades not violate spatially differentiated deposition standards, the solution to the joint cost minimization problem in general is not attainable through a series of sequential and bilateral permit trades. The reason is that the solution to a sequential and bilateral trading process is path dependent; that is, the sequence of bilateral trades determines the outcome. A particular sequence of trades would lead deposition at a particular geographical location to increase up to the maximum constrained level at that location, while a different sequence of trades would lead a different deposition constraint to bind first. Once a deposition constraint binds, it is not always possible for further cost-saving bilateral trades to be individually consummated without violating the deposition constraint, even though additional trades *taken as a set* may lower costs while complying with all deposition constraints.

At least four studies have shown that in the context of SO_2 emissions trading in Europe a sequential trading process will result in savings compared to CAC, but fall far short of the potential savings at the cost minimum, due to the problem of path dependence (Klaassen and Amann, 1992; Kruitwagen, 1992; van Ierland *et al.*, 1993; Klaassen and Førsund, 1993). A consistent assumption in these studies is that trading takes place between nations rather than between individual enterprises. The assumption is necessitated by the use of the RAINS model as a platform for the simulations. Cost functions in the RAINS model are aggregated at the national level. The presumption in the model is that abatement efforts *within* a nation will be pursued in a mostly cost-effective manner, although some explicit prohibitions are built in.

In the program of research reported in this chapter, this assumption is relaxed in order to explore the performance of bilateral and sequential trading algorithms when trading occurs at the enterprise level rather than at the level of national governments. A trading simulation is constructed using piecewise linear cost functions drawn from RAINS for various small sets of European countries presumed to enter an international system of permit trading to reduce SO_2 emissions. Nations in each set are selected arbitrarily, with some effort made to ensure geographic proximity. Transport coefficients between countries are drawn from data for 1990 from EMEP (Sandnes, 1993). This toy model is reconstructed repeatedly using different sets of nations to test the robustness of the results. The performance of the algorithm for one scenario is presented in *Figure 12.1* and explained below.

Trading is modeled as a bilateral process with a stochastic element identifying the sequence in which potential trades are considered. Before each trade, potential trading partners are chosen using a Monte Carlo sampling procedure. Differences in marginal costs provide an incentive to engage in trade, so intuition suggests that agents with the greatest difference in marginal costs would be more likely to engage in trade than agents with relatively similar marginal costs. However, there are a variety of factors

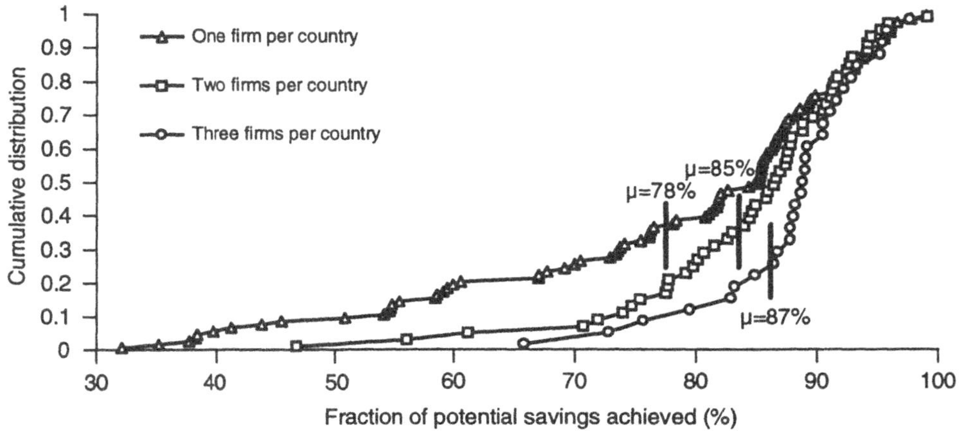

Figure 12.1. Savings from three levels of disaggregation.

that also seem important in identifying trading partners, including an element of chance in what may be characterized as a matching and bargaining process. These missing factors are represented by associating probabilistic weights with each pair of potential trading partners. The weights may be varied along a continuum, with differences in marginal costs playing a deterministic role in identifying potential trading partners at one extreme, and at the other extreme allowing the order of trades that are explored to be entirely random. We represent this continuum for a pair of potential traders i and j are by the weight

$$w_{ij} = |mc_i - mc_j|^{\ln(1/\theta)} / \sum_i \sum_{j>i} |mc_i - mc_j|^{\ln(1/\theta)}, 0 \leq \theta \leq 1, \qquad (12.1)$$

where θ is a parameter that reflects the amount of randomness in the determination of trading partners. In the example represented in *Figure 12.1*, θ is assigned the value 1, implying that potential pairs are identified randomly. That is, each pair i and j has an equal probability of being selected.

After a potential pair of trading partners is selected, the direction of possible trade is determined by checking the relative magnitude in marginal costs, and the ability to trade is determined by checking deposition constraints for all nations. If trade is possible, permits are exchanged until marginal costs are equal or until a deposition constraint binds. Whether trade is possible or not, the selected pair of potential traders remain eligible to be selected again in the next pairing. In the simulations presented in *Figure 12.1*, there is a 1% possibility the process will conclude after each potential pairing is considered. (Other stopping rules are also considered, but are not discussed here.) After trading concludes, the outcome is evaluated according to the percentage of potential gains from trade that are achieved. Potential gains from trade are calibrated by the difference between global compliance costs at the joint cost minimization solution (equivalently, multilateral simultaneous trading) and the cost if no trading occurs. Finally, the simulation is repeated, each time identifying a potentially unique sequence of trades and resulting outcome.

In *Figure 12.1*, the horizontal axis represents the percentage of potential savings that are achieved. Each simulation yields a different outcome, and these outcomes are organized as a cumulative distribution function along this dimension. The vertical axis represents probability, and the three curves represent different cumulative distribution functions for three comparable scenarios. One scenario involved trading between six national governments. The nations in this example included Belgium, France, Italy, the Netherlands, Spain, and the UK.

A second scenario represents trading between 12 representative enterprises, two located in each of the six countries. To represent an enterprise, each nations cost function was apportioned randomly to constitute two large firms. In the third scenario, the same procedure was followed to represent enterprise level trading by three large firms in each country.

In this experiment, enterprise trading outperformed trading by national governments. This is represented in *Figure 12.1*, for example, by the placement of the curve for trading with two firms per nation to the right of the curve for trading directly between nations. This indicates that the expected percentage of the potential savings that were realized across repeated simulations was greater when there were two firms per nation. Furthermore, the more disaggregated the identity of the trading agents, the greater the percentage of potential savings that were realized. For instance, in *Figure 12.1* the curve for trading with three firms per nation lies to the right of the curve for trading with two firms per nation. The mean percent of savings that were achieved under each scenario is indicated on the graph as μ. But perhaps more interesting is the shape of the overall distributions. An approximate first-order stochastic dominance is achieved by each level of disaggregation in the agents designated for trade.

A systematic discussion of this simulation experiment, covering a variety of national groupings and with sensitivity analysis, is presented in Burtraw *et al.* (1993). A number of other scenarios combining various groupings of nations and employing different assumptions have obtained results similar to those in *Figure 12.1*, but they also indicate sensitivity to a number of assumptions about how the process of bilateral trading would proceed – that is, how the sequence of potential trades is selected – and to how the national cost functions are disaggregated to achieve representative enterprises. However, under most assumptions these simulations support the finding that bilateral and sequential trading at the disaggregated enterprise level outperform trading between national governments in terms of the cost savings that can be expected. The results for the simulation presented here are not meant to be definitive or conclusive, but are presented for illustrative purposes only. They are suggestive of results that may obtain in a fuller analysis of these issues.

12.3 Comparing Performance in Promoting Allocative Efficiency

12.3.1 Introduction

The second potential benefit attributed to IB environmental regulation is the internalization of social costs in the product market, thereby promoting allocative efficiency. In a system of tradable permits, the social opportunity cost of emissions is reflected in the market price of a permit. If the system works ideally, this price is internalized in private financial decisions along with other costs such as investments in pollution abatement. A tradable permit system is distinguished from CAC regulation because the latter approach internalizes only the costs of abatement, not the social cost of residual emissions.

To illustrate this, consider a hypothetical example in which the government has full information about costs and emissions. In this case, CAC regulation could achieve a cost-effective assignment of pollution abatement by codifying in technology design standards the same outcome that would be obtained by an IB approach, where presumably the marginal cost of abatement would be equal to the marginal environmental benefit of emission reduction. However, even in this case, sometimes referred to as "smart" CAC regulation, the social cost of residual pollution will not be reflected in the price of a factor input or final product.

The internalization of social costs in product prices, including the price of electricity, is critical to the environmental and economic objectives of the EU. The 1987 Single European Act and the 1991 Treaty on Political Union place environmental protection on an equal footing with economic growth, the free movement of goods and services, and economic competition in the

Table 12.1. The internalization of social cost under command-and-control regulation.

	Clean technology	Dirty technology (unabated)	Dirty technology (with abatement)
Private costs of electricity production	10	7	7
Private costs of pollution abatement			2
External cost of residual pollution		5	2
Total private financial costs	10	7	9
Total social costs	10	12	11

policies of the EU. However, it is not widely appreciated that environmental protection and the form that it takes is essential to achieving these other economic objectives. The internalization of social costs is critical to economic integration because social costs, including environmental costs that are not internalized, constitute an implicit hidden subsidy to a given technology or industry.

This point is illustrated in *Table 12.1*, where the first column represents the marginal cost for producing electricity with a hypothetical "clean" technology that has no externalities. The second column represents a "dirty" technology that is unabated, and the third column represents the dirty technology under smart CAC regulation. Total social costs are greater for the dirty technology, even when it is regulated by smart CAC, than they are for the clean technology. If regulation of the environment was incentive-based, the technology with least social cost would be selected. However, private investment decisions made under a smart CAC policy would choose the abated dirty technology, leading to a misallocation of resources from the social perspective.

Subsidies have played an historic role in European energy markets. For instance, since 1965 aid to the coal industry may have cost European taxpayers more than 70 billion ECUs. This estimate includes only fiscal subsidies and does not include hidden subsidies associated with the effects of pollution (CEC, 1991; Royal Institute of International Affairs/Science Policy Research Unit, 1989). A goal of the "polluter-pays" principle is to discourage member states of the EU from granting subsidies to polluters (Coudert Brothers, 1992, p. 8). Subsidies yield anticompetitive advantage to selected enterprises and the elimination of subsidies is recognized as a necessary precondition for economic unification in the EU. For these reasons, the internalization of social costs is a fundamental precept to the attainment of both economic and environmental goals in the EU. These goals would not be achieved through cost-effective CAC policies. But more important for this discussion, the incentive to perpetuate existing subsidies may permeate the implementation of an IB approach to environmental regulation. The next section considers how the designation of agency may be important to whether the continuation of subsidies is possible.

12.3.2 Strategic considerations in implementation

Consider an international agreement for SO_2 emission control that specifies (potentially tradable) emission targets for each country, but leaves the mechanism for achieving targets up to the individual country. Hoel (1991b) discusses the incentive for unilateral action in the setting of an environmental goal, but he does not address the incentive for adopting different mechanisms to achieve that goal. Similarly, Ottinger, (1993) discusses the incentive for competitors within the OECD not to unilaterally adopt a carbon tax (see also CEC, 1992). Previously I asked whether there is reason to believe that, acting unilaterally, national governments would implement national systems of tradable emission permits. The suggestion was that even in the face of an external emission cap imposed on a nation, the considerable inertia associated with CAC approaches to regulation would be likely to keep it the approach used in the design of domestic environmental policy. A second reason to question whether a nation would use IB approaches to comply with an exogenous emission cap stems from the potential for strategic behavior in the design of environmental policy, and the effect that different types of environmental policy have on competitiveness in an international open economy.

Imagine that a nation acting unilaterally was to adopt an emissions trading program to comply with an internationally negotiated emission cap, or a cap achieved through international permit trading between national governments. This approach to domestic policy would be expected to generate savings in compliance cost expenditures compared to a CAC approach. For instance, in England and Wales, the cost of complying with the Large Combustion Plant Directive in the year 2003 with tradable permits could be reduced by 13%, or about £45–£65 million a year in 1990 prices (London Economics, 1992, pp. 23–24, 77). However, the national government must weigh these savings against the detriment to international competitiveness from including external costs in product prices (Bovenberg, 1993). As indicated in *Table 12.1*, a consequence of IB regulation is likely to be higher

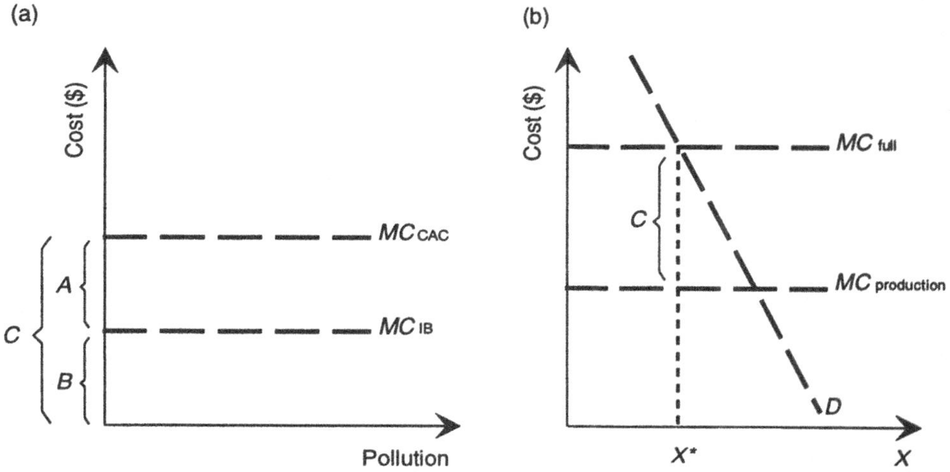

Figure 12.2. Marginal cost of pollution control and the product market.

product prices because those prices would include not only the marginal cost of pollution control but also the opportunity cost of using the atmosphere's absorptive capacity. In *Table 12.1*, the total private cost of the clean technology is 10; this is greater than the private cost of 9, but less than the total social cost of 11 for the abated dirty technology. The distinction between marginal and average costs makes a direct comparison difficult, but in general it is probable that the opportunity cost of permit holdings will constitute more than the cost savings from IB regulation.

The illustration in *Figure 12.2* may help make this idea more precise. In *Figure 12.2(a)* the marginal cost of pollution control for a hypothetical industry is represented; for simplicity it is assumed to be constant. The cost-effective marginal cost of control is labeled MC_{IB}, with a measure of B. The marginal cost of control with CAC regulation is labeled MC_{CAC}, with a measure of C. The difference between these two costs is labeled A.

In *Figure 12.2(b)* characteristics of the product market are depicted when CAC regulation is in place. The constant marginal cost of production for the industry, absent pollution abatement, is labeled $MC_{\text{production}}$. The marginal cost of pollution control C is added to this to indicate the industry's full cost. Demand is labeled D, and x^* indicates the quantity of output.

Consider a shift to the use of tradable permits. We assume this shift would cause a dramatic 50% reduction in the cost of pollution control for given output x^*, that is, $A = B$. However, in the product market the reduction in abatement cost would be just offset by the cost of permits. If the permit market is efficient, the price of a permit would equal the marginal cost of abatement, or B. Assuming competitive markets, this opportunity cost would be recognized by the industry even if permits were endowed at zero cost. The total cost of environmental regulation would be $2B = C$. Consequently, the abatement cost savings would be just offset by the opportunity cost of permits, with no change in the marginal operating cost of the industry or in industry output.

In comparison, imagine a less dramatic reduction in cost in shifting from CAC to tradable permits, so that $A < B$. Holding output constant at x^*, the total cost of environmental regulation on the industry would increase from $A + B$ to $2B$. Output would subsequently fall. Of course, if cost savings from the use of tradable permits were much greater, output could expand, but few advocates for IB regulation make claims quite this bold. Consequently, in the typical case the costs for the industry would be expected to rise, at least from a partial equilibrium perspective.

A concrete example may be found in examining the US trading program for SO_2 allowances. The annual opportunity cost of permit holdings are expected to be about \$2.6 billion, based on an assumption that permits will have a market value of \$300 per ton of SO_2, compared to a potential annual compliance cost under the system of tradable permits of perhaps \$3.5 billion. The actual market value of permits in the USA remains uncertain, but it is likely to be between \$200 and \$300. The potential savings in compliance costs of the trading program compared to a CAC approach may be as high as \$1.5 billion. This estimate is closely related to the price of permits in a complicated manner that includes developments in secondary markets related to compliance planning. In any case, it is fair to assume that a lower permit price would indicate that these cost savings would be lower as well. Hence, if the opportunity cost of permit holdings is fully recognized by the industry and is treated as an expense (something that is unlikely to occur fully due to the nature of the electricity industry) the trading program could be viewed as raising prices in the industry by \$1.1 billion (\$2.6 billion less \$1.5 billion). If a CAC approach had been substituted, US society might be able to subsidize electricity prices by \$2.6 billion with monetary expenditures of \$1.5 billion. The remainder of the subsidy takes the form of allocative inefficiency.

The opportunity to reduce product prices with an implicit subsidy of the magnitude in this example suggests that the benefits to an industry of a CAC approach may greatly outweigh the cost savings of using incentive-based regulation. The benefits of the implicit subsidy will be distributed unevenly within a nations economy, providing rewards for those who utilize polluting technology and penalizing those who utilize cleaner technology. In the European setting, individual nations are more interdependent economically than the USA is with its trading partners, and consequently the national benefits of competitive pricing may be even larger.

To illustrate the potential situation in Europe, we attempted to calculate the aggregate asset value of all permit holdings at an equilibrium in a tradable SO_2 permit market, where spatially differentiated deposition constraints are in effect. To identify an equilibrium, or optimal allocation of permits, we evaluated a scenario for a 30% uniform reduction in SO_2 emissions from the Official Energy Pathway by all nations excluding republics of the former Soviet Union using the RAINS model. We then took the deposition pattern at this level of emissions as constraints for an emissions trading scenario. In other words, we assumed that permits were allocated or were traded so as to just satisfy the constraints implied by this deposition pattern. A model of the transportation of emissions to deposition was based on data for 1990 (Sandnes, 1993).

Our goal was to calculate the marginal opportunity cost of each ton emitted and permit used. This calculation is more delicate than the example for the USA above, because of the assumed spatially differentiated deposition constraints. We calculated the maximum amount that each country i would receive from the sale of a single permit to any other country j. Since country i would reduce its emissions through a sale and country j would increase its emissions, the net effect on binding deposition constraints must be taken into account and in many cases one permit from country i would allow country j to increase its emissions by less than one permit equivalent. Therefore for each country i we searched across all possible trading partners to find a potential buyer j of a marginal permit from i that would pay more than any other country for that permit. The maximum amount j would pay for a permit from i is assumed to be j's marginal abatement cost multiplied by the ratio of allowable increases in emissions by j, subject to the deposition constraints, if i reduced emissions by one unit. In general, at the equilibrium, the opportunity cost of a unit of emissions will be less than the marginal cost of emission reduction for i or it would be possible to trade a permit and reduce cost further.

The annual opportunity cost for permit use is summed for all countries and estimated to be DM3.68 billion. Under an assumption of competitive markets, this sum would be internalized in the price of goods and services and would have a sizable effect on the competitive standing of some nations, and some industries in particular. The incentive to avoid this impact on prices and to retain the implicit subsidy for polluting industries associated with CAC regulation is clear, and would be acutely felt by national governments saddled with responsibility to design domestic environmental policies that comply with a national emission cap. Admittedly, recent strategic trade theory (stemming from Brander and Spencer, 1985; and Eaton and Grossman, 1986) suggests there are only rare instances in which fiscal subsidies such as this would make sense, so it can not be assumed that it would be rational for a national government to institute such a subsidy in the future. However, the subsidy that is described here is one that takes the form of allocative inefficiency that is generally unpriced and which does not enter the government's budget constraint. Furthermore, the political economy of removing a subsidy that already is in place makes unilateral changes in policy even more problematic.

Hence, in the absence of a specific mechanism to implement incentive-based environmental regulation as a component of international environmental agreements, it may not be rational for national governments unilaterally to adopt such an approach in Europe. The potential consequences are twofold. One is the increase in cost due to the failure to capture the potential cost savings. The second is an undermining of the necessary precepts for economic integration in Europe: the elimination of subsidies implicitly provided by some national environmental policies and the maintenance of a level playing field for trade.

12.4 A Perspective on the Role of National Governments

Another characteristic of IB approaches to environmental regulation is that the government's role in economic activities typically is reduced compared to CAC regulation, because the government is relieved of having to set technical standards for production processes. This alleged virtue has nothing to do with ideological convictions about the proper role of government. In fact, on a practical level this is sometimes cited to bolster a position in favor of more stringent environmental standards because those standards can be achieved with less direct government interference in business activities, and therefore, in principle, at less cost.

Some argue that a lessened role for government in the operations of industrial facilities is offset by an increased role in monitoring, since the government can not rely on the installation of specific pollution control technologies to predict environmental compliance by firms as would typify a CAC approach. However, experience has taught that the installation of specific technologies does not guarantee predicted compliance under CAC regulation without commensurate investments in monitoring. Often control equipment is expensive to operate, and its performance requires regular investments in maintenance that may not occur in a CAC setting in the absence of government supervision.

It is also sometimes suggested that IB regulation may impart a greater incentive for firms to cheat on compliance because the firm not only saves on its abatement costs, as it would under a CAC regime, but it saves on the cost of permits, taxes, or other relevant policy instruments. But for the same reason, IB approaches invite agents to monitor other agents with regard to compliance because the competitive effects of undetected cheating are greater, and because instruments such as tradable permits have an asset value for a potential monitoring agent that is affected by the behavior of other firms. It is often suggested that IB approaches incorporate provisions for legal standing by third parties, including nongovernmental organizations, and a structure of fines that will encourage private parties to complement governmental efforts in monitoring and enforcement. Nonetheless, on balance, one must conclude that in many cases a greater level of involvement by government in monitoring and enforcement may be more appropriate with IB approaches than with the CAC alternative.

However, in the case of regulating SO_2 emissions, it is noteworthy that new continuous emission monitoring systems (CEMS) allow regulators to obtain precise information about emissions at electricity generating facilities through measurement at the smokestack, in the absence of information about the fuels and processes used. The 1990 US Clean Air Act amendments implementing tradable SO_2 emission allowances rely heavily on this technology for monitoring. Hence, the degree of government involvement necessary to perform a rigorous monitoring function may be eased considerably.

Both types of governmental activities – he specification of technology-based standards in the operation of enterprises, and the monitoring of compliance activities – require familiarity with the industrial processes at individual facilities. Similarly, whether trading occurs between national governments or between firms, a similar role for national governments in monitoring is implied. If national governments engage in trade, they must enforce domestic regulations, just as they would presumably be assigned the primary role in enforcing the obligations of firms in an international trading regime.

The issues examined in this chapter lead me to suggest that if governments are the trading agents in a system of tradable permits there is a greater opportunity for inefficient implementation than if enterprises assume that role. Resting entitlements to permits and authority for compliance planning in the hands of national governments, which are open to the influence of competing social priorities and private interests, is unlikely to lead to the widespread adoption of IB environmental policies at the domestic level. Current regulatory policies grounded in CAC approaches are likely to change only slowly, in the absence of an impetus from the international community. Furthermore, national governments are becoming increasingly aware that domestic industries must compete in an increasingly competitive global economy. Consequently, this conclusion seems especially valid where there is an incentive noncompatibility between national interests, or the interests of important constituencies, and the use of IB approaches. These same conclusions apply *a fortiori* to the electric utility industry, which is the most important sector with regard to SO_2 emissions.

This chapter has considered just two economic criteria – cost effectiveness and the internalization of social costs. One might readily suggest that other criteria are also important for the design of social policy. Among the most important of these are social concerns about the welfare of constituencies affected by dramatic changes in social policy. There may be compelling reasons to preserve the current design of environmental policy, or at least to ease the transition to more efficient approaches. The purpose here is simply to consider the incentives and anticipated performance of the two alternative approaches to agency according to rather narrow criteria.

Whether an international permit system with enterprises acting as the trading agents would succeed further according to these criteria remains an open question. A number of regulatory issues will likely intervene to undermine emissions trading in any case, as discussed in the previous sections. However, under a system in which firms engage directly in permit trading it appears that firms would at least be more likely to recognize the opportunity cost for marginal acquisitions of permits, which would not occur if they were regulated solely by CAC regulations under domestic statutes recalibrated to comply with international agreements. Furthermore, an important point to note is that placing permits directly into the hands of affected enterprises removes one obvious mechanism for perpetuating current subsidies in the form of nonpriced allocative inefficiencies.

Perhaps the most far-reaching consequence of this problem does not relate to the quality of the environment. At least in the short run, environmental assets would be protected equally well whether nations or enterprises are the trading agents in a European SO_2 accord. Of course, in the long run, international support for environmental goals will depend on the cost of attainment, and the prospects for environmental improvement rest in large part with the prospects for technological innovation, which are improved if permit trading is effective. But the ultimate measure of the success of international environmental policies in Europe may lie in their influence on the process of economic unification in the EU and economic integration with central and eastern Europe.

Acknowledgments

Financial support for this research was provided in part by the US Environmental Protection Agency. The author is grateful to David Edelstein, Ken Harrison, Ger Klaassen, Paul Turner and three anonymous referees for assistance and guidance.

References

Atkinson, S., and Tietenberg, T., 1991, Market Failure in Incentive-Based Regulation: The Case of Emissions Trading, *Journal of Environmental Economics and Management* **21**:17–31.

Baumol, W.J., and Oates, W.E., 1988, *The Theory of Environmental Policy*, 2nd ed., Cambridge University Press, New York, NY, USA.

Bertram, G., 1992, Tradable Emission Permits and the Control of Greenhouse Gases, *Journal of Development Studies* **28**(3):423–446.

Bohi, D.R., and Burtraw, D., 1991, Avoiding Regulatory Gridlock in the Acid Rain Program, *Journal of Policy Analysis and Management* **10**(4):676–684.

Bohi, D.R., and Burtraw, D., 1992, Utility Investment Behavior and the Emissions Trading Market, *Resources and Energy* **14**:129–153.

Bohm, P., 1981, *Deposit-Refund Systems: Theory and Applications to Environmental, Conservation, and Consumer Policy*, published for Resources for the Future, Inc. by The Johns Hopkins University Press, Baltimore, MD, USA.

Bovenberg, A.L., 1993, Policy Instruments for Curbing CO_2 Emissions: The Case of the Netherlands, *Environmental and Resource Economics* **3**(3):233–244.

Brander, J.A., and Spencer, B.J., 1985, Export Subsidies and International Market Share Rivalry, *Journal of International Economics* **18**:83–100.

Burtraw, D., Harrison, K., and Turner, P., 1993, *Path Dependence in Bilateral Emissions Trading*, Working Paper, Resources for the Future, Washington, DC, USA.

CEC (Commission of the European Communities), 1991, Interview with Commissioner António Cardoso e Cunha, in *Energy in Europe: Energy Policies and Trends in the European Community*, Directorate General for Energy, Number 18 (December).

CEC (Commission of the European Communities), 1992, *Proposal for a Council Directive Introducing a Tax on Carbon Dioxide Emissions and Energy*, COM [92] 226 final, Brussels, Belgium.

Coudert Brothers, 1992, *The Environmental Policy of the European Community*, Brussels, Belgium.

Dudek, D.J., and Palmisano, J., 1988, Emissions Trading: Why Is This Thoroughbred Hobbled? *Columbia Journal of Environmental Law* **13**:217–256.

Dudek, D.J., and Tietenberg, T.H., 1992, Monitoring and Enforcing Greenhouse Gas Trading, in *Climate Change: Designing a Tradable Permit System*, OECD, Paris, France.

Eaton, J., and Grossman, G.M., 1986, Optimal Trade and Industrial Policy under Oligopoly, *Quarterly Journal of Economics* **191**:383–406.

Grubb, M., 1992, Options for an International Agreement, in *Combating Global Warming*, UN, New York, NY, USA.

Hahn, R.W., 1984, Market Power and Transferable Property Rights, *Quarterly Journal of Economics* **xcix**(4):753–765.

Hahn, R.W., 1989, Economic Prescriptions for Environmental Problems: How the Patient Followed the Doctor's Orders, *Journal of Economic Perspectives* **3**(2):95–114.

Hahn, R.W., and Hester, G.L., 1989, Where Did All the Markets Go? An Analysis of EPA's Emissions Trading Program, *Yale Journal on Regulation* **6**:109–153.

Hausker, K., 1992, The Politics and Economics of Auction Design in the Market for Sulfur Dioxide Pollution, *Journal of Policy Analysis and Management* **11**(4):553–572.

Henney, A., 1992, *The Electricity Supply Industries of Eleven West European Countries*, Energy Economic Engineering Limited and European Energy Economics, London, UK.

Hoel, M., 1991a, Efficient International Agreements for Reducing Emissions of CO_2, *The Energy Journal* **12**(2):93–107.

Hoel, M., 1991b, Global Environmental Problems: The Effects of Unilateral Actions Taken by One Country, *Journal of Environmental Economics and Management* **20**:55–70.

Hoel, M., 1993, Harmonization of Carbon Taxes in International Climate Agreements, *Environmental and Resource Economics* **3**(3):221–231.

Holmes, A., 1988, *Electricity in Europe: Opening the Market*, Financial Times Business Information, London, UK.

Kete, N., 1992, *The Politics of Markets: The Acid Rain Control Policy in the 1990 Clean Air Act Amendments*, Ph.D. Dissertation, The Johns Hopkins University, Baltimore, MD, USA.

Klaassen, G., and Amann, M., 1992, *Trading of Emission Reduction Commitments for Sulfur Dioxide in Europe*, SR-92-03, International Institute for Applied Systems Analysis, Laxenburg, Austria.

Klaassen, G.A.J., and Førsund, F.R., 1993, *Emission Trading in Europe with an Exchange Rate*, Paper prepared for the Fourth Annual Conference of the European Association of Environmental and Resource Economists, June 30–July 3, Fontainebleau, France.

Kruitwagen, S., 1992, *Tradable Permits for SO_2-Emissions in Europe: A Pollution Offset Experiment*, Wageningen Economic Papers, Wageningen Agricultural University, The Netherlands.

London Economics, 1992, *The Potential Role of Market Mechanisms in the Control of Acid Rain*, HMSO, Department of The Environment, Environmental Economics Research Series, London, UK.

Malueg, D.A., 1990, Welfare Consequences of Emission Credit Trading Programs, *Journal of Environmental Economics and Management* **18**:66–77.

Misiolek, W., and Elder, H., 1989, Exclusionary Manipulation of Markets for Pollution Rights, *Journal of Environmental Economics and Management* **16**:156–166.

Montgomery, W.D., 1972, Markets in Licenses and Efficient Pollution Control Programs, *Journal of Economic Theory* **5**:395–418.

OECD (Organisation for Economic Co-operation and Development), 1992, *Climate Change: Designing a Tradable Permit System*, Paris, France.

OECD (Organisation for Economic Co-operation and Development), 1993, *Taxation and the Environment: Complementary Policies*, Paris, France.

Ottinger, R.L., 1993, Pollution Taxes: The Preferred Means of Incorporation of Environmental Externalities, *Proceedings of a Conference on Incorporation of Environmental Externalities in Resource Acquisition Decisions in the United States and Europe*, September 8–11, Wingspread Conference Center, Racine, WI, Springer-Verlag, Berlin, Germany.

Royal Institute of International Affairs/Science Policy Research Unit, 1989, *A Single European Market in Energy*, Energy and Environment Programme, London, UK.

Sandnes, H., 1993, Corrigendum to EMEP/MSC-w Report 1/92 (Appendix C), EMEP/MSC-W Report 1/92, The Norwegian Meteorological Institute.

Shaw, R., Alcamo, J., and Hordijk, L., 1990, Strategy Development and Assessment Using RAINS," in J. Alcamo, R. Shaw, and L. Hordijk, eds., *The RAINS Model of Acidification*, Kluwer Academic Publishers, Dordrecht, The Netherlands.

Swart, R., 1992, Greenhouse Gas Emissions Trading: Defining the Commodity, in *Climate Change: Designing a Tradable Permit System*, OECD, Paris, France.

Tietenberg, T.H., 1985, *Emissions Trading: An Exercise in Reforming Pollution Policy*, Resources for the Future, Washington, DC, USA.

Tietenberg, T.H., 1990, *The Implementation and Effectiveness of Emissions Trading in the US*, Department of Economics Working Paper, Colby College, Waterville, ME, USA.

Tripp, J.T.B., and Dudek, D.J., 1989, Institutional Guidelines for Designing Successful Transferable Rights Programs, *Yale Journal on Regulation* **6**:369–391.

UN, 1992, *Combating Global Warming*, United Nations, New York, NY, USA.

UN ECE (United Nations Economic Commission for Europe), 1992, *Impacts of Long-range Transboundary Air Pollution*, Air Pollution Studies No. 8, United Nations, New York, NY, USA.

van Ierland, E.C., Kruitwagen, S., and Hendrix, E.M.T., 1993, *Tradable Discharge Permits for Acidifying Emissions in Europe: "Guided" Bilateral Trade*, Department of Economics, Wageningen Agricultural University, The Netherlands.

Index

Economy & Environment

1. F. Archibugi and P. Nijkamp (eds.): *Economy and Ecology: Towards Sustainable Development.* 1989 ISBN 0-7923-0477-2
2. J. Bojö, K.-G. Mäler and L. Unemo: *Environment and Development: An Economic Approach.* 1990 ISBN 0-7923-0802-6
3. J. B. Opschoor and D. W. Pearce (eds.): *Persistent Pollutants: Economics and Policy.* 1991 ISBN 0-7923-1168-X
4. D.J. Kraan and R. J. in 't Veld (eds.): *Environmental Protection: Public or Private Choice.* 1991 ISBN 0-7923-1333-X
5. J.J. Krabbe and W.J.M. Heijman (eds.): *National Income and Nature: Externalities, Growth and Steady State.* 1992 ISBN 0-7923-1529-4
6. J. Bojö, K.-G. Mäler and L. Unemo: *Environment and Development: An Economic Approach* (revised edition). 1992 ISBN 0-7923-1878-1
7. T. Sterner (ed.): *Economic Policies for Sustainable Development.* 1994 ISBN 0-7923-2680-6
8. L. Bergman and D.M. Pugh (eds.): *Environmental Toxicology, Economics and Institutions.* The Atrazine Case Study. 1994 ISBN 0-7923-2986-4
9. G. Klaassen and F.R. Førsund (eds.): *Economic Instruments for Air Pollution Control.* 1994 ISBN 0-7923-3151-6

KLUWER ACADEMIC PUBLISHERS – DORDRECHT / BOSTON / LONDON

The manufacturer's authorised representative in the EU is Springer Nature Customer Service Centre GmbH, Europaplatz 3, 69115 Heidelberg, Germany. If you have any concerns regarding our products, please contact ProductSafety@springernature.com

Printed and bound by CPI Group (UK) Ltd, Croydon, CR0 4YY
10/10/2026
02231227-0006